Student Solutions Manual
Volume 2
for
Tipler and Mosca's
Physics for Scientists and Engineers
Sixth Edition

David Mills
Professor Emeritus
College of the Redwoods

W. H. Freeman and Company
New York

© 2008 by W. H. Freeman and Company

Printed in the United States of America

ISBN-13: 978-1-4292-0303-6
ISBN-10: 1-4292-0303-X (Volume 2: Chapters 21–33)

First printing

W. H. Freeman and Company
41 Madison Avenue
New York, NY 10010
Houndmills, Basingstoke
RG21 6XS, England
www.whfreeman.com

Contents

Acknowledgments

Anthony J. Buffa (professor emeritus of California Polytechnic State University, San Luis Obispo) and Todd Pedlar (Luther College) provided the new problems appearing in the Sixth Edition. Tony saved me many hours of work by providing rough-draft solutions to these new problems and collaborated in clarifying the physics of the solutions. Gene Mosca (formerly of the United States Naval Academy and co-author of the Sixth Edition) also helped clarify many of my solutions and provided guidance when I was unsure how best to proceed. It was a pleasure to collaborate with both Tony and Gene in the creation of this solutions manual. They share my hope that you will find these solutions useful in learning physics.

Michael Crivello (San Diego Mesa College) and Carlos Delgado (Community College of Southern Nevada) checked the solutions. Without their thorough work, many errors would have remained to be discovered by the users of this solutions manual. Carlos also suggested several alternate solutions, all of which were improvements on mine, and they are included in the solutions manual. Their assistance is greatly appreciated. In spite of their best efforts, there may still be errors in some of the solutions, and for those I assume full responsibility. Should you find errors or think of alternative solutions that you would like to call to my attention, please do not hesitate to send them to me by using asktipler@whfreeman.com.

It was a pleasure to work with Susan Brennan, Clancy Marshall, and Kharissia Pettus who guided us through the creation of this solutions manual. I would also like to thank Danielle Storm and Janie Chan for organizing the reviewing process.

April 2007

David Mills
Professor Emeritus
College of the Redwoods

To the Student

This solution manual accompanies *Physics for Scientists and Engineers, 6e,* by Paul Tipler and Gene Mosca. Following the structure of the solutions to the worked examples in the text, we begin a solution to an end-of-chapter numerical problem by picturing the problem—representing the problem pictorially whenever appropriate and expressing the physics of the solution in the form of a mathematical model. Then, the problem is solved or any intermediate steps are filled in as needed, the appropriate substitutions and algebraic simplifications are made, and the solution with the substitution of numerical values (including their units) is completed. This problem-solving strategy is used by experienced learners of physics and it is our hope that you will see the value in such an approach to problem solving and learn to use it consistently.

Believing that it will maximize your learning of physics, we encourage you to create your own solution before referring to the solutions in this manual. You may find that, by following this approach, you will find different, but equally valid, solutions to some of the problems. In any event, studying the solutions contained herein without having first attempted the problems will do little to help you learn physics.

Chapter 21
The Electric Field I: Discrete Charge Distributions

Conceptual Problems

13 ••   Two point particles that have charges of $+q$ and $-3q$ are separated by distance d. (*a*) Use field lines to sketch the electric field in the neighborhood of this system. (*b*) Draw the field lines at distances much greater than d from the charges.

Determine the Concept (*a*) We can use the rules for drawing electric field lines to draw the electric field lines for this system. In the field-line sketch we've assigned 2 field lines to each charge q. (*b*) At distances much greater than the separation distance between the two charges, the system of two charged bodies will "look like" a single charge of $-2q$ and the field pattern will be that due to a point charge of $-2q$. Four field lines have been assigned to each charge $-q$.

(*a*)   (*b*)

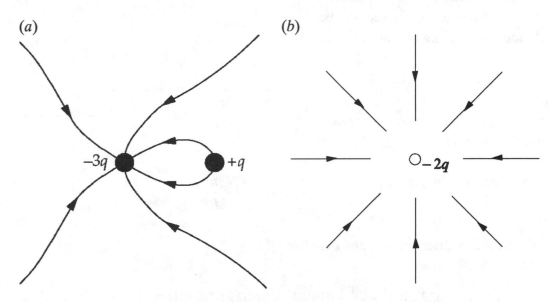

17 •••   Two molecules have dipole moments of equal magnitude. The dipole moments are oriented in various configurations as shown in Figure 21-34. Determine the electric-field direction at each of the numbered locations. Explain your answers.

Determine the Concept Figure 21-23 shows the electric field due to a single dipole, where the dipole moment is directed toward the right. The electric field due two a pair of dipoles can be obtained by superposing the two electric fields.

	1	2	3
(*a*)	down	up	up
(*b*)	up	right	left

$$(c) \quad \text{down} \quad \text{up} \quad \text{up}$$
$$(d) \quad \text{down} \quad \text{up} \quad \text{up}$$

Charge

23 •   What is the total charge of all of the protons in 1.00 kg of carbon?

Picture the Problem We can find the number of coulombs of positive charge there are in 1.00 kg of carbon from $Q = 6n_C e$, where n_C is the number of atoms in 1.00 kg of carbon and the factor of 6 is present to account for the presence of 6 protons in each atom. We can find the number of atoms in 1.00 kg of carbon by setting up a proportion relating Avogadro's number, the mass of carbon, and the molecular mass of carbon to n_C. See Appendix C for the molar mass of carbon.

Express the positive charge in terms of the electronic charge, the number of protons per atom, and the number of atoms in 1.00 kg of carbon:

$$Q = 6n_C e$$

Using a proportion, relate the number of atoms in 1.00 kg of carbon n_C, to Avogadro's number and the molecular mass M of carbon:

$$\frac{n_C}{N_A} = \frac{m_C}{M} \Rightarrow n_C = \frac{N_A m_C}{M}$$

Substitute for n_C to obtain:

$$Q = \frac{6N_A m_C e}{M}$$

Substitute numerical values and evaluate Q:

$$Q = \frac{6\left(6.022 \times 10^{23} \frac{\text{atoms}}{\text{mol}}\right)(1.00\,\text{kg})\left(1.602 \times 10^{-19}\,\text{C}\right)}{0.01201 \frac{\text{kg}}{\text{mol}}} = \boxed{4.82 \times 10^7\,\text{C}}$$

Coulomb's Law

27 •   Three point charges are on the x-axis: $q_1 = -6.0\,\mu\text{C}$ is at $x = -3.0$ m, $q_2 = 4.0\,\mu\text{C}$ is at the origin, and $q_3 = -6.0\,\mu\text{C}$ is at $x = 3.0$ m. Find the electric force on q_1.

Picture the Problem q_2 exerts an attractive electric force $\vec{F}_{2,1}$ on point charge q_1 and q_3 exerts a repulsive electric force $\vec{F}_{3,1}$ on point charge q_1. We can find the net electric force on q_1 by adding these forces (that is, by using the superposition principle).

Express the net force acting on q_1:

$$\vec{F}_1 = \vec{F}_{2,1} + \vec{F}_{3,1}$$

Express the force that q_2 exerts on q_1:

$$\vec{F}_{2,1} = \frac{k|q_1||q_2|}{r_{2,1}^2}\hat{i}$$

Express the force that q_3 exerts on q_1:

$$\vec{F}_{3,1} = \frac{k|q_1||q_3|}{r_{3,1}^2}\left(-\hat{i}\right)$$

Substitute and simplify to obtain:

$$\vec{F}_1 = \frac{k|q_1||q_2|}{r_{2,1}^2}\hat{i} - \frac{k|q_1||q_3|}{r_{3,1}^2}\hat{i}$$

$$= k|q_1|\left(\frac{|q_2|}{r_{2,1}^2} - \frac{|q_3|}{r_{3,1}^2}\right)\hat{i}$$

Substitute numerical values and evaluate $\vec{F}_1$:

$$\vec{F}_1 = \left(8.988\times10^9\ \text{N}\cdot\text{m}^2/\text{C}^2\right)\left(6.0\ \mu\text{C}\right)\left(\frac{4.0\ \mu\text{C}}{(3.0\ \text{m})^2} - \frac{6.0\ \mu\text{C}}{(6.0\ \text{m})^2}\right)\hat{i} = \boxed{\left(1.5\times10^{-2}\ \text{N}\right)\hat{i}}$$

35 ••• Five identical point charges, each having charge Q, are equally spaced on a semicircle of radius R as shown in Figure 21-37. Find the force (in terms of k, Q, and R) on a charge q located equidistant from the five other charges.

Picture the Problem By considering the symmetry of the array of charged point particles, we can see that the y component of the force on q is zero. We can apply Coulomb's law and the principle of superposition of forces to find the net force acting on q.

Express the net force acting on the point charge q:

$$\vec{F}_q = \vec{F}_{Q\,\text{on}\,x\,\text{axis},q} + 2\vec{F}_{Q\,\text{at}\,45°,q}$$

Express the force on point charge q due to the point charge Q on the x axis:

$$\vec{F}_{Q \, on \, x \, axis, q} = \frac{kqQ}{R^2}\hat{i}$$

Express the net force on point charge q due to the point charges at 45°:

$$2\vec{F}_{Q \, at \, 45°, q} = 2\frac{kqQ}{R^2}\cos 45°\hat{i}$$

$$= \frac{2}{\sqrt{2}}\frac{kqQ}{R^2}\hat{i}$$

Substitute for $\vec{F}_{Q \, on \, x \, axis, q}$ and $2\vec{F}_{Q \, at \, 45°, q}$ to obtain:

$$\vec{F}_q = \frac{kqQ}{R^2}\hat{i} + \frac{2}{\sqrt{2}}\frac{kqQ}{R^2}\hat{i}$$

$$= \boxed{\frac{kqQ}{R^2}\left(1 + \sqrt{2}\right)\hat{i}}$$

The Electric Field

37 •   A point charge of 4.0 μC is at the origin. What is the magnitude and direction of the electric field on the x axis at (*a*) $x = 6.0$ m, and (*b*) $x = -10$ m? (*c*) Sketch the function E_x versus x for both positive and negative values of x. (Remember that E_x is negative when $\vec{E}$ points in the $-x$ direction.)

Picture the Problem Let q represent the point charge at the origin and use Coulomb's law for $\vec{E}$ due to a point charge to find the electric field at $x = 6.0$ m and -10 m.

(*a*) Express the electric field at a point P located a distance x from a point charge q:

$$\vec{E}(x) = \frac{kq}{x^2}\hat{r}_{P,0}$$

Evaluate this expression for $x = 6.0$ m:

$$\vec{E}(6.0\,m) = \frac{\left(8.988 \times 10^9 \, \frac{N \cdot m^2}{C^2}\right)(4.0\,\mu C)}{(6.0\,m)^2}\hat{i} = \boxed{(0.10\,kN/C)\hat{i}}$$

(*b*) Evaluate $\vec{E}$ at $x = -10$ m:

$$\vec{E}(-10\,m) = \frac{\left(8.988 \times 10^9 \, \frac{N \cdot m^2}{C^2}\right)(4.0\,\mu C)}{(10\,m)^2}(-\hat{i}) = \boxed{(-0.36\,kN/C)\hat{i}}$$

(c) The following graph was plotted using a spreadsheet program:

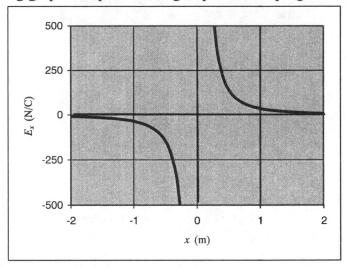

41 •• Two point charges q_1 and q_2 both have a charge equal to +6.0 nC and are on the y axis at $y_1 = +3.0$ cm and $y_2 = -3.0$ cm respectively. (a) What is the magnitude and direction of the electric field on the x axis at $x = 4.0$ cm? (b) What is the force exerted on a third charge $q_0 = 2.0$ nC when it is placed on the x axis at $x = 4.0$ cm?

Picture the Problem The diagram shows the locations of the point charges q_1 and q_2 and the point on the x axis at which we are to find $\vec{E}$. From symmetry considerations we can conclude that the y component of $\vec{E}$ at any point on the x axis is zero. We can use Coulomb's law for the electric field due to point charges and the principle of superposition for fields to find the field at any point on the x axis and $\vec{F} = q\vec{E}$ to find the force on a point charge q_0 placed on the x axis at $x = 4.0$ cm.

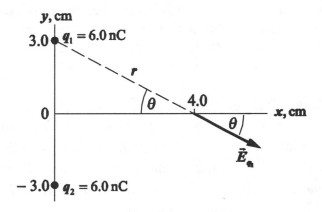

(*a*) Letting $q = q_1 = q_2$, express the *x*-component of the electric field due to one point charge as a function of the distance *r* from either point charge to the point of interest:

$$\vec{E}_x = \frac{kq}{r^2}\cos\theta\,\hat{i}$$

Express $\vec{E}_x$ for both charges:

$$\vec{E}_x = 2\frac{kq}{r^2}\cos\theta\,\hat{i}$$

Substitute for $\cos\theta$ and *r*, substitute numerical values, and evaluate to obtain:

$$\vec{E}(4.0\,\text{cm})_x = 2\frac{kq}{r^2}\frac{0.040\,\text{m}}{r}\hat{i} = \frac{2kq(0.040\,\text{m})}{r^3}\hat{i}$$

$$= \frac{2(8.988\times10^9\,\text{N}\cdot\text{m}^2/\text{C}^2)(6.0\,\text{nC})(0.040\,\text{m})}{\left[(0.030\,\text{m})^2 + (0.040\,\text{m})^2\right]^{3/2}}\hat{i}$$

$$= (34.5\,\text{kN/C})\hat{i} = (35\,\text{kN/C})\hat{i}$$

The magnitude and direction of the electric field at *x* = 4.0 cm is:

$$\boxed{35\ \text{kN/C @ 0°}}$$

(*b*) Apply $\vec{F} = q\vec{E}$ to find the force on a point charge q_0 placed on the *x* axis at *x* = 4.0 cm:

$$\vec{F} = (2.0\,\text{nC})(34.5\,\text{kN/C})\hat{i}$$

$$= \boxed{(69\,\mu\text{N})\hat{i}}$$

47 •• Two point particles, each having a charge *q*, sit on the base of an equilateral triangle that has sides of length *L* as shown in Figure 21-38. A third point particle that has a charge equal to 2*q* sits at the apex of the triangle. Where must a fourth point particle that has a charge equal to *q* be placed in order that the electric field at the center of the triangle be zero? (The center is in the plane of the triangle and equidistant from the three vertices.)

Picture the Problem The electric field of 4th charged point particle must cancel the sum of the electric fields due to the other three charged point particles. By symmetry, the position of the 4th charged point particle must lie on the vertical centerline of the triangle. Using trigonometry, one can show that the center of an equilateral triangle is a distance $L/\sqrt{3}$ from each vertex, where *L* is the length of the side of the triangle. Note that the *x* components of the fields due to the base charged particles cancel each other, so we only need concern ourselves with the *y* components of the fields due to the charged point particles at the vertices of the triangle. Choose a coordinate system in which the origin is at the midpoint of the base of the triangle, the +*x* direction is to the right, and the +*y* direction is upward.

Note that the x components of the electric field vectors add up to zero.

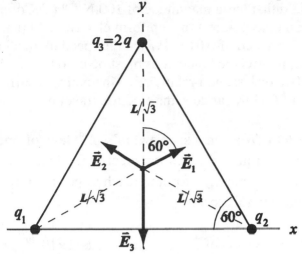

Express the condition that must be satisfied if the electric field at the center of the triangle is to be zero:

$$\sum_{i=1 \text{ to } 4} \vec{E}_i = 0$$

Substituting for $\vec{E}_1$, $\vec{E}_2$, $\vec{E}_3$, and $\vec{E}_4$ yields:

$$\frac{k(q)}{\left(\dfrac{L}{\sqrt{3}}\right)^2}\cos 60°\,\hat{j} + \frac{k(q)}{\left(\dfrac{L}{\sqrt{3}}\right)^2}\cos 60°\,\hat{j} - \frac{k(2q)}{\left(\dfrac{L}{\sqrt{3}}\right)^2}\,\hat{j} + \frac{kq}{y^2}\,\hat{j} = 0$$

Solving for y yields:

$$y = \pm\frac{L}{\sqrt{3}}$$

Because the positive solution corresponds to the 4th charge being at the center of the triangle, it follows that:

$$y = -\frac{L}{\sqrt{3}}$$

The charge must be placed a distance equal to the length of the side of the triangle divided by the square root of three below the midpoint of the base of the triangle, where L is the length of a side of the triangle.

Point Charges in Electric Fields

51 ••   The acceleration of a particle in an electric field depends on q/m (the charge-to-mass ratio of the particle). (*a*) Compute q/m for an electron.

(*b*) What is the magnitude and direction of the acceleration of an electron in a uniform electric field that has a magnitude of 100 N/C? (*c*) Compute the time it takes for an electron placed at rest in a uniform electric field that has a magnitude of 100 N/C to reach a speed of 0.01*c*. (When the speed of an electron approaches the speed of light *c*, relativistic kinematics must be used to calculate its motion, but at speeds of 0.01*c* or less, non-relativistic kinematics is sufficiently accurate for most purposes.) (*d*) How far does the electron travel in that time?

Picture the Problem We can use Newton's 2nd law of motion to find the acceleration of the electron in the uniform electric field and constant-acceleration equations to find the time required for it to reach a speed of 0.01*c* and the distance it travels while acquiring this speed.

(*a*) Use data found at the back of your text to compute *e*/*m* for an electron:

$$\frac{e}{m_e} = \frac{1.602 \times 10^{-19}\,\text{C}}{9.109 \times 10^{-31}\,\text{kg}}$$

$$= \boxed{1.76 \times 10^{11}\,\text{C/kg}}$$

(*b*) Apply Newton's 2nd law to relate the acceleration of the electron to the electric field:

$$a = \frac{F_{\text{net}}}{m_e} = \frac{eE}{m_e}$$

Substitute numerical values and evaluate *a*:

$$a = \frac{\left(1.602 \times 10^{-19}\,\text{C}\right)\left(100\,\text{N/C}\right)}{9.109 \times 10^{-31}\,\text{kg}}$$

$$= 1.759 \times 10^{13}\,\text{m/s}^2$$

$$= \boxed{1.76 \times 10^{13}\,\text{m/s}^2}$$

The direction of the acceleration of an electron is opposite the electric field.

(*c*) Using the definition of acceleration, relate the time required for an electron to reach 0.01*c* to its acceleration:

$$\Delta t = \frac{v}{a} = \frac{0.01c}{a}$$

Substitute numerical values and evaluate Δt:

$$\Delta t = \frac{0.01\left(2.998 \times 10^8\,\text{m/s}\right)}{1.759 \times 10^{13}\,\text{m/s}^2} = 0.1704\,\mu s$$

$$= \boxed{0.2\,\mu s}$$

(d) Find the distance the electron travels from its average speed and the elapsed time:

$$\Delta x = v_{av}\Delta t = \tfrac{1}{2}\left[0 + 0.01\left(2.998\times10^8 \text{ m/s}\right)\right]\left(0.1704\,\mu s\right) = \boxed{3\,\text{mm}}$$

57 ••   An electron starts at the position shown in Figure 21-39 with an initial speed $v_0 = 5.00 \times 10^6$ m/s at 45° to the x axis. The electric field is in the +y direction and has a magnitude of 3.50×10^3 N/C. The black lines in the figure are charged metal plates. On which plate and at what location will the electron strike?

Picture the Problem We can use constant-acceleration equations to express the x and y coordinates of the electron in terms of the parameter t and Newton's 2nd law to express the constant acceleration in terms of the electric field. Eliminating the parameter will yield an equation for y as a function of x, q, and m. We can decide whether the electron will strike the upper plate by finding the maximum value of its y coordinate. Should we find that it does not strike the upper plate, we can determine where it strikes the lower plate by setting $y(x) = 0$. Ignore any effects of gravitational forces.

Express the x and y coordinates of the electron as functions of time:	$x = \left(v_0 \cos\theta\right)t$ and $y = \left(v_0 \sin\theta\right)t - \tfrac{1}{2}a_y t^2$
Apply Newton's 2nd law to relate the acceleration of the electron to the net force acting on it:	$a_y = \dfrac{F_{net,y}}{m_e} = \dfrac{eE_y}{m_e}$
Substitute in the y-coordinate equation to obtain:	$y = \left(v_0 \sin\theta\right)t - \dfrac{eE_y}{2m_e}t^2$
Eliminate the parameter t between the two equations to obtain:	$y(x) = \left(\tan\theta\right)x - \dfrac{eE_y}{2m_e v_0^2 \cos^2\theta}x^2$ (1)
To find y_{max}, set $dy/dx = 0$ for extrema:	$\dfrac{dy}{dx} = \tan\theta - \dfrac{eE_y}{m_e v_0^2 \cos^2\theta}x'$ $= 0$ for extrema
Solve for x′ to obtain:	$x' = \dfrac{m_e v_0^2 \sin 2\theta}{2eE_y}$ (See remark below.)

Substitute x' in $y(x)$ and simplify to obtain y_{max}:

$$y_{max} = \frac{m_e v_0^2 \sin^2 \theta}{2eE_y}$$

Substitute numerical values and evaluate y_{max}:

$$y_{max} = \frac{(9.109 \times 10^{-31}\ kg)(5.00 \times 10^6\ m/s)^2 \sin^2 45°}{2(1.602 \times 10^{-19}\ C)(3.50 \times 10^3\ N/C)} = 1.02\ cm$$

and, because the plates are separated by 2 cm, the electron does not strike the upper plate.

To determine where the electron will strike the lower plate, set $y = 0$ in equation (1) and solve for x to obtain:

$$x = \frac{m_e v_0^2 \sin 2\theta}{eE_y}$$

Substitute numerical values and evaluate x:

$$x = \frac{(9.109 \times 10^{-31}\ kg)(5.00 \times 10^6\ m/s)^2 \sin 90°}{(1.602 \times 10^{-19}\ C)(3.50 \times 10^3\ N/C)} = \boxed{4.1\ cm}$$

Remarks: x' is an extremum, that is, either a maximum or a minimum. To show that it is a maximum we need to show that d^2y/dx^2, evaluated at x', is negative. A simple alternative is to use your graphing calculator to show that the graph of $y(x)$ is a maximum at x'. Yet another alternative is to recognize that, because equation (1) is quadratic and the coefficient of x^2 is negative, its graph is a parabola that opens downward.

General Problems

61 • Show that it is only possible to place one isolated proton in an ordinary empty coffee cup by considering the following situation. Assume the first proton is fixed at the bottom of the cup. Determine the distance directly above this proton where a second proton would be in equilibrium. Compare this distance to the depth of an ordinary coffee cup to complete the argument.

Picture the Problem Equilibrium of the second proton requires that the sum of the electric and gravitational forces acting on it be zero. Let the upward direction be the +y direction and apply the condition for equilibrium to the second proton.

Apply $\sum F_y = 0$ to the second proton:

$$\vec{F}_e + \vec{F}_g = 0$$

or

$$\frac{kq_p^2}{h^2} - m_p g = 0 \Rightarrow h = \sqrt{\frac{kq_p^2}{m_p g}}$$

Substitute numerical values and evaluate h:

$$h = \sqrt{\frac{\left(8.988 \times 10^9 \ \frac{N \cdot m^2}{C^2}\right)\left(1.602 \times 10^{-19} \ C\right)^2}{\left(1.673 \times 10^{-27} \ kg\right)\left(9.81 \ m/s^2\right)}} \approx 12 \ cm \approx 5 \ in$$

This separation of about 5 in is greater than the height of a typical coffee cup. Thus the first proton will repel the second one out of the cup and the maximum number of protons in the cup is one.

65 •• A positive charge Q is to be divided into two positive point charges q_1 and q_2. Show that, for a given separation D, the force exerted by one charge on the other is greatest if $q_1 = q_2 = \frac{1}{2}Q$.

Picture the Problem We can use Coulomb's law to express the force exerted on one charge by the other and then set the derivative of this expression equal to zero to find the distribution of the charge that maximizes this force.

Using Coulomb's law, express the force that either charge exerts on the other:

$$F = \frac{kq_1q_2}{D^2}$$

Express q_2 in terms of Q and q_1:

$$q_2 = Q - q_1$$

Substitute for q_2 to obtain:

$$F = \frac{kq_1(Q - q_1)}{D^2}$$

Differentiate F with respect to q_1 and set this derivative equal to zero for extreme values:

$$\frac{dF}{dq_1} = \frac{k}{D^2}\frac{d}{dq_1}[q_1(Q - q_1)]$$

$$= \frac{k}{D^2}[q_1(-1) + Q - q_1]$$

$$= 0 \ \text{for extrema}$$

Solve for q_1 to obtain:

$$q_1 = \frac{1}{2}Q \Rightarrow q_2 = Q - q_1 = \frac{1}{2}Q$$

To determine whether a maximum or a minimum exists at $q_1 = \frac{1}{2}Q$, differentiate F a second time and evaluate this derivative at $q_1 = \frac{1}{2}Q$:

$$\frac{d^2F}{dq_1^2} = \frac{k}{D^2}\frac{d}{dq_1}[Q - 2q_1]$$

$$= \frac{k}{D^2}(-2)$$

< 0 independently of q_1.

$$\boxed{\therefore q_1 = q_2 = \tfrac{1}{2}Q \text{ maximizes } F.}$$

69 •• A rigid 1.00-m-long rod is pivoted about its center (Figure 21-42). A charge $q_1 = 5.00 \times 10^{-7}$ C is placed on one end of the rod, and a charge $q_2 = -q_1$ is placed a distance $d = 10.0$ cm directly below it. (*a*) What is the force exerted by q_2 on q_1? (*b*) What is the torque (measured about the rotation axis) due to that force? (*c*) To counterbalance the attraction between the two charges, we hang a block 25.0 cm from the pivot as shown. What value should we choose for the mass of the block? (*d*) We now move the block and hang it a distance of 25.0 cm from the balance point, on the same side of the balance as the charge. Keeping q_1 the same, and d the same, what value should we choose for q_2 to keep this apparatus in balance?

Picture the Problem We can use Coulomb's law, the definition of torque, and the condition for rotational equilibrium to find the electrostatic force between the two charged bodies, the torque this force produces about an axis through the center of the rod, and the mass required to maintain equilibrium when it is located either 25.0 cm to the right or to the left of the mid-point of the rod.

(*a*) Using Coulomb's law, express the electric force between the two charges:

$$F = \frac{kq_1q_2}{d^2}$$

Substitute numerical values and evaluate F:

$$F = \frac{(8.988 \times 10^9 \text{ N} \cdot \text{m}^2/\text{C}^2)(5.00 \times 10^{-7} \text{ C})^2}{(0.100\,\text{m})^2} = 0.2247 \text{ N} = \boxed{0.225 \text{ N}}$$

(*b*) Apply the definition of torque to obtain:

$$\tau = F\ell$$

Substitute numerical values and evaluate τ:

$$\tau = (0.2247 \text{ N})(0.500\,\text{m})$$

$$= 0.1124 \text{ N} \cdot \text{m} = \boxed{0.112 \text{ N} \cdot \text{m}},$$

counterclockwise.

(*c*) Apply $\sum \tau_{\substack{\text{center of} \\ \text{the rod}}} = 0$ to the rod:

$$\tau - mg\ell' = 0 \Rightarrow m = \frac{\tau}{g\ell'}$$

Substitute numerical values and evaluate m:

$$m = \frac{0.1124\,\text{N}\cdot\text{m}}{(9.81\,\text{m/s}^2)(0.250\,\text{m})}$$

$$= 0.04582\,\text{kg} = \boxed{45.8\,\text{g}}$$

(d) Apply $\sum \tau_{\substack{\text{center of}\\\text{the rod}}} = 0$ to the rod:

$$-\tau + mg\ell' = 0$$

Substitute for τ.

$$-F\ell + mg\ell' = 0$$

Substitute for F:

$$-\frac{kq_1q_2'}{d^2} + mg\ell' = 0 \Rightarrow q_2' = \frac{d^2 mg\ell'}{kq_1\ell}$$

where q' is the required charge.

Substitute numerical values and evaluate q_2':

$$q_2' = \frac{(0.100\,\text{m})^2(0.04582\,\text{kg})(9.81\,\text{m/s}^2)(0.250\,\text{m})}{(8.988\times10^9\,\text{N}\cdot\text{m}^2/\text{C}^2)(5.00\times10^{-7}\,\text{C})(0.500\,\text{m})} = \boxed{5.00\times10^{-7}\,\text{C}}$$

71 ••   Two point charges have a total charge of 200 μC and are separated by 0.600 m. (a) Find the charge of each particle if the particles repel each other with a force of 120 N. (b) Find the force on each particle if the charge on each particle is 100 μC.

Picture the Problem Let the numeral 1 denote one of the small spheres and the numeral 2 the other. Knowing the total charge on the two spheres, we can use Coulomb's law to find the charge on each of them. A second application of Coulomb's law when the spheres carry the same charge and are 0.600 m apart will yield the force each exerts on the other.

(a) Use Coulomb's law to express the repulsive force each charge exerts on the other:

$$F = \frac{kq_1q_2}{r_{1,2}^2}$$

Express q_2 in terms of the total charge and q_1:

$$q_2 = Q - q_1$$

Substitute for q_2 to obtain:

$$F = \frac{kq_1(Q-q_1)}{r_{1,2}^2}$$

Substitute numerical values to obtain:

$$120\,\text{N} = \frac{\left(8.988\times10^{9}\ \text{N}\cdot\text{m}^{2}/\text{C}^{2}\right)\left[(200\,\mu\text{C})q_{1}-q_{1}^{2}\right]}{(0.600\,\text{m})^{2}}$$

Simplify to obtain the quadratic equation:

$$q_{1}^{2}+\left(-200\,\mu\text{C}\right)q_{1}+4805\left(\mu\text{C}\right)^{2}=0$$

Use the quadratic formula or your graphing calculator to obtain:

$$q_{1}=28.0\,\mu\text{C and }172\,\mu\text{C}$$

Hence the charges on the particles are:

$$\boxed{28.0\,\mu\text{C}}\text{ and }\boxed{172\,\mu\text{C}}$$

(b) Use Coulomb's law to express the repulsive force each charge exerts on the other when $q_1 = q_2 = 100\ \mu C$:

$$F = \frac{kq_{1}q_{2}}{r_{1,2}^{2}}$$

Substitute numerical values and evaluate F:

$$F = \left(8.988\times10^{9}\ \text{N}\cdot\text{m}^{2}/\text{C}^{2}\right)\frac{(100\,\mu\text{C})^{2}}{(0.600\,\text{m})^{2}} = \boxed{250\,\text{N}}$$

77 ••   Figure 21-46 shows a dumbbell consisting of two identical small particles, each of mass m, attached to the ends of a thin (massless) rod of length a that is pivoted at its center. The particles carry charges of $+q$ and $-q$, and the dumbbell is located in a uniform electric field $\vec{E}$. Show that for small values of the angle θ between the direction of the dipole and the direction of the electric field, the system displays a rotational form of simple harmonic motion, and obtain an expression for the period of that motion.

Picture the Problem We can apply Newton's 2nd law in rotational form to obtain the differential equation of motion of the dipole and then use the small angle approximation $\sin\theta \approx \theta$ to show that the dipole experiences a linear restoring torque and, hence, will experience simple harmonic motion.

Apply $\sum \tau = I\alpha$ to the dipole:

$$-pE\sin\theta = I\frac{d^{2}\theta}{dt^{2}}$$

where τ is negative because acts in such a direction as to decrease θ.

For small values of θ, $\sin\theta \approx \theta$ and:

$$-pE\theta = I\frac{d^2\theta}{dt^2}$$

Express the moment of inertia of the dipole:

$$I = \tfrac{1}{2}ma^2$$

Relate the dipole moment of the dipole to its charge and the charge separation:

$$p = qa$$

Substitute for p and I to obtain:

$$\tfrac{1}{2}ma^2\frac{d^2\theta}{dt^2} = -qaE\theta$$

or

$$\boxed{\frac{d^2\theta}{dt^2} = -\frac{2qE}{ma}\theta}$$

the differential equation for a simple harmonic oscillator with angular frequency $\omega = \sqrt{2qE/ma}$.

Express the period of a simple harmonic oscillator:

$$T = \frac{2\pi}{\omega}$$

Substitute for ω and simplify to obtain:

$$T = \boxed{2\pi\sqrt{\frac{ma}{2qE}}}$$

79 •• An electron (charge $-e$, mass m) and a positron (charge $+e$, mass m) revolve around their common center of mass under the influence of their attractive coulomb force. Find the speed v of each particle in terms of e, m, k, and their separation distance L.

Picture the Problem The forces the electron and the proton exert on each other constitute an action-and-reaction pair. Because the magnitudes of their charges are equal and their masses are the same, we find the speed of each particle by finding the speed of either one. We'll apply Coulomb's force law for point charges and Newton's 2nd law to relate v to e, m, k, and their separation distance L.

Apply Newton's 2nd law to the positron to obtain:

$$\frac{ke^2}{L^2} = m\frac{v^2}{\tfrac{1}{2}L} \Rightarrow \frac{ke^2}{L} = 2mv^2$$

Solve for v to obtain:

$$v = \sqrt{\frac{ke^2}{2mL}}$$

85 •••   During a famous experiment in 1919, Ernest Rutherford shot doubly ionized helium nuclei (also known as alpha particles) at a gold foil. He discovered that virtually all of the mass of an atom resides in an extremely compact nucleus.   Suppose that during such an experiment, an alpha particle far from the foil has an initial kinetic energy of 5.0 MeV. If the alpha particle is aimed directly at the gold nucleus, and the only force acting on it is the electric force of repulsion exerted on it by the gold nucleus, how close will it approach the gold nucleus before turning back? That is, what is the minimum center-to-center separation of the alpha particle and the gold nucleus?

Picture the Problem The work done by the electric field of the gold nucleus changes the kinetic energy of the alpha particle–eventually bringing it to rest. We can apply the work-kinetic energy theorem to derive an expression for the distance of closest approach. Because the repulsive Coulomb force $\vec{F}_e$ varies with distance, we'll have to evaluate $\int \vec{F}_e \cdot d\vec{r}$ in order to find the work done on the alpha particles by this force.

Apply the work-kinetic energy theorem to the alpha particle to obtain:

$$W_{net} = \int_{\infty}^{r_{min}} \vec{F}_e \cdot d\vec{r} = \Delta K$$

or, because

$$\int_{\infty}^{r_{min}} \vec{F}_e \cdot d\vec{r} = -\int_{\infty}^{r_{min}} \frac{k(2e)(79e)}{r^2} dr \text{ and}$$

$K_f = 0$,

$$-158ke^2 \int_{\infty}^{r_{min}} \frac{dr}{r^2} = -K_i$$

Evaluating the integral yields:

$$-158ke^2 \left[\frac{1}{r}\right]_{\infty}^{r_{min}} = -\frac{158ke^2}{r_{min}} = -K_i$$

Solve for r_{min} and simplify to obtain:

$$r_{min} = \frac{158ke^2}{K_i}$$

Substitute numerical values and evaluate r_{min}:

$$r_{min} = \frac{158\left(8.988\times10^9\ \frac{N\cdot m^2}{C^2}\right)\left(1.602\times10^{-19}C\right)^2}{5.0\ Mev\times\dfrac{1.602\times10^{-19}\ J}{eV}} = \boxed{4.6\times10^{-14}\ m}$$

87 •••   In Problem 86, there is a description of the Millikan experiment used to determine the charge on the electron. During the experiment, a switch is used to reverse the direction of the electric field without changing its magnitude, so that one can measure the terminal speed of the microsphere both as it is moving upward and as it is moving downward. Let v_u represent the terminal speed when the particle is moving up, and v_d the terminal speed when moving down. (a) If we let $u = v_u + v_d$, show that $q = 3\pi\eta ru/E$, where q is the microsphere's net charge. For the purpose of determining q, what advantage does measuring both v_u and v_d have over measuring only one terminal speed? (b) Because charge is quantized, u can only change by steps of magnitude N, where N is an integer. Using the data from Problem 86, calculate Δu.

Picture the Problem The free body diagram shows the forces acting on the microsphere of mass m and having an excess charge of $q = Ne$ when the electric field is downward. Under terminal-speed conditions the sphere is in equilibrium under the influence of the electric force $\vec{F}_e$, its weight $m\vec{g}$, and the drag force $\vec{F}_d$. We can apply Newton's 2nd law, under terminal-speed conditions, to relate the number of excess charges N on the sphere to its mass and, using Stokes' law, to its terminal speed.

(a) Apply Newton's 2nd law to the microsphere when the electric field is downward:

$F_e - mg - F_d = ma_y$
or, because $a_y = 0$,
$F_e - mg - F_{d,terminal} = 0$

Substitute for F_e and $F_{d,terminal}$ to obtain:

$qE - mg - 6\pi\eta rv_u = 0$
or, because $q = Ne$,
$NeE - mg - 6\pi\eta rv_u = 0$

Solve for v_{u} to obtain:

$$v_{\mathrm{u}} = \frac{NeE - mg}{6\pi\eta r} \qquad (1)$$

With the field pointing upward, the electric force is downward and the application of Newton's 2nd law to the microsphere yields:

$$F_{d,\,\mathrm{terminal}} - F_e - mg = 0$$
or
$$6\pi\eta r v_{\mathrm{d}} - NeE - mg = 0$$

Solve for v_{d} to obtain:

$$v_{\mathrm{d}} = \frac{NeE + mg}{6\pi\eta r} \qquad (2)$$

Add equations (1) and (2) and simplify to obtain:

$$u = v_{\mathrm{u}} + v_{\mathrm{d}} = \frac{NeE - mg}{6\pi\eta\, r} + \frac{NeE + mg}{6\pi\eta\, r} = \frac{NeE}{3\pi\eta\, r} = \boxed{\frac{qE}{3\pi\eta\, r}}$$

 This has the advantage that you don't need to know the mass of the microsphere.

(b) Letting Δu represent the change in the terminal speed of the microsphere due to a gain (or loss) of one electron we have:

$$\Delta u = v_{N+1} - v_N$$

Noting that Δv will be the same whether the microsphere is moving upward or downward, express its terminal speed when it is moving upward with N electronic charges on it:

$$v_N = \frac{NeE - mg}{6\pi\eta r}$$

Express its terminal speed upward when it has $N + 1$ electronic charges:

$$v_{N+1} = \frac{(N+1)eE - mg}{6\pi\eta r}$$

Substitute and simplify to obtain:

$$\Delta u = \frac{(N+1)eE - mg}{6\pi\eta r} - \frac{NeE - mg}{6\pi\eta r}$$

$$= \frac{eE}{6\pi\eta r}$$

Substitute numerical values and evaluate Δu:

$$\Delta u = \frac{\left(1.602 \times 10^{-19}\ \mathrm{C}\right)\left(6.00 \times 10^4\ \mathrm{N/C}\right)}{6\pi\left(1.8 \times 10^{-5}\ \mathrm{Pa \cdot m}\right)\left(5.50 \times 10^{-7}\ \mathrm{m}\right)}$$

$$= \boxed{52\ \mu\mathrm{m/s}}$$

Chapter 22
The Electric Field II: Continuous Charge Distributions

Conceptual Problems

1   •   Figure 22-37 shows an L-shaped object that has sides which are equal in length. Positive charge is distributed uniformly along the length of the object. What is the direction of the electric field along the dashed 45° line? Explain your answer.

Determine the Concept The resultant field is directed along the dashed line; pointing away from the intersection of the two sides of the L-shaped object. This can be seen by dividing each leg of the object into 10 (or more) equal segments and then drawing the electric field on the dashed line due to the charges on each pair of segments that are equidistant from the intersection of the legs.

7   ••   An electric dipole is completely inside a closed imaginary surface and there are no other charges. True or False:

(*a*)   The electric field is zero everywhere on the surface.
(*b*)   The electric field is normal to the surface everywhere on the surface.
(*c*)   The electric flux through the surface is zero.
(*d*)   The electric flux through the surface could be positive or negative.
(*e*)   The electric flux through a portion of the surface might not be zero.

(*a*) False. Near the positive end of the dipole, the electric field, in accordance with Coulomb's law, will be directed outward and will be nonzero. Near the negative end of the dipole, the electric field, in accordance with Coulomb's law, will be directed inward and will be nonzero.

(*b*) False. The electric field is perpendicular to the Gaussian surface only at the intersections of the surface with a line defined by the axis of the dipole.

(*c*) True. Because the net charge enclosed by the Gaussian surface is zero, the net flux, given by $\phi_{net} = \oint_S E_n dA = 4\pi k Q_{inside}$, through this surface must be zero.

(*d*) False. The flux through the closed surface is zero.

(*e*) True. All Gauss's law tells us is that, because the net charge inside the surface is zero, the *net* flux through the surface must be zero.

9 ••   Suppose that the total charge on the conducting spherical shell in Figure 22-38 is zero. The negative point charge at the center has a magnitude given by Q. What is the direction of the electric field in the following regions? (*a*) $r < R_1$, (*b*) $R_2 > r > R_1$, (*c*) and $r > R_2$. Explain your answer.

Determine the Concept We can apply Gauss's law to determine the electric field for $r < R_1$, $R_2 > r > R_1$, and $r > R_2$. We also know that the direction of an electric field at any point is determined by the direction of the electric force acting on a positively charged object located at that point.

(*a*) From the application of Gauss's law we know that the electric field in this region is not zero. A positively charged object placed in the region for which $r < R_1$ will experience an attractive force from the charge $-Q$ located at the center of the shell. Hence the direction of the electric field is radially inward.

(*b*) Because the total charge on the conducting sphere is zero, the charge on its inner surface must be positive (the positive charges in the conducting sphere are drawn there by the negative charge at the center of the shell) and the charge on its outer surface must be negative. Hence the electric field in the region $R_2 > r > R_1$ is radially outward.

(*c*) Because the charge on the outer surface of the conducting shell is negative, the electric field in the region $r > R_2$ is radially inward.

Calculating $\vec{E}$ From Coulomb's Law

13 ••   A uniform line charge that has a linear charge density l equal to 3.5 nC/m is on the x axis between $x = 0$ and $x = 5.0$ m. (*a*) What is its total charge? Find the electric field on the x axis at (*b*) $x = 6.0$ m, (*c*) $x = 9.0$ m, and (*d*) $x = 250$ m.
(*e*) Estimate the electric field at $x = 250$ m, using the approximation that the charge is a point charge on the x axis at $x = 2.5$ m, and compare your result with the result calculated in Part (*d*). (To do this you will need to assume that the values given in this problem statement are valid to more than two significant figures.) Is your approximate result greater or smaller than the exact result? Explain your answer.

Picture the Problem We can use the definition of λ to find the total charge of the line of charge and the expression for the electric field on the axis of a finite line of charge to evaluate E_x at the given locations along the x axis. In Part (*d*) we can apply Coulomb's law for the electric field due to a point charge to approximate the electric field at $x = 250$ m.

(a) Use the definition of linear charge density to express Q in terms of λ:

$$Q = \lambda L = (3.5\,\text{nC/m})(5.0\,\text{m}) = 17.5\,\text{nC}$$
$$= \boxed{18\,\text{nC}}$$

Express the electric field on the axis of a finite line charge:

$$E_x(x_0) = \frac{kQ}{x_0(x_0 - L)}$$

(b) Substitute numerical values and evaluate E_x at $x = 6.0$ m:

$$E_x(6.0\,\text{m}) = \frac{(8.988\times10^9\,\text{N}\cdot\text{m}^2/\text{C}^2)(17.5\,\text{nC})}{(6.0\,\text{m})(6.0\,\text{m} - 5.0\,\text{m})} = \boxed{26\,\text{N/C}}$$

(c) Substitute numerical values and evaluate E_x at $x = 9.0$ m:

$$E_x(9.0\,\text{m}) = \frac{(8.988\times10^9\,\text{N}\cdot\text{m}^2/\text{C}^2)(17.5\,\text{nC})}{(9.0\,\text{m})(9.0\,\text{m} - 5.0\,\text{m})} = \boxed{4.4\,\text{N/C}}$$

(d) Substitute numerical values and evaluate E_x at $x = 250$ m:

$$E_x(250\,\text{m}) = \frac{(8.988\times10^9\,\text{N}\cdot\text{m}^2/\text{C}^2)(17.5\,\text{nC})}{(250\,\text{m})(250\,\text{m} - 5.0\,\text{m})} = 2.56800\ \text{mN/C} = \boxed{2.6\,\text{mN/C}}$$

(e) Use Coulomb's law for the electric field due to a point charge to obtain:

$$E_x(x) = \frac{kQ}{x^2}$$

Substitute numerical values and evaluate $E_x(250$ m):

$$E_x(250\,\text{m}) = \frac{(8.988\times10^9\,\text{N}\cdot\text{m}^2/\text{C}^2)(17.5\,\text{nC})}{(250\,\text{m} - 2.5\ \text{m})^2} = 2.56774\,\text{mN/C} = \boxed{2.6\,\text{mN/C}}$$

This result is about 0.01% less than the exact value obtained in (d). This suggests that the line of charge is too long for its field at a distance of 250 m to be modeled exactly as that due to a point charge.

17 •   A ring that has radius a lies in the $z = 0$ plane with its center at the origin. The ring is uniformly charged and has a total charge Q. Find E_z on the z axis at (a) $z = 0.2a$, (b) $z = 0.5a$, (c) $z = 0.7a$, (d) $z = a$, and (e) $z = 2a$. (f) Use your results to plot E_z versus z for both positive and negative values of z. (Assume that these distances are exact.)

Picture the Problem We can use $E_z = 2\pi kq\left(1 - \dfrac{z}{\sqrt{z^2 + a^2}}\right)$ to find the electric

field at the given distances from the center of the charged ring.

(*a*) Evaluate $E_z(0.2a)$:

$$E_z(0.2a) = \frac{kQ(0.2a)}{\left[(0.2a)^2 + a^2\right]^{3/2}}$$

$$= \boxed{0.189 \frac{kQ}{a^2}}$$

(*b*) Evaluate $E_z(0.5a)$:

$$E_z(0.5a) = \frac{kQ(0.5a)}{\left[(0.5a)^2 + a^2\right]^{3/2}}$$

$$= \boxed{0.358 \frac{kQ}{a^2}}$$

(*c*) Evaluate $E_z(0.7a)$:

$$E_z(0.7a) = \frac{kQ(0.7a)}{\left[(0.7a)^2 + a^2\right]^{3/2}}$$

$$= \boxed{0.385 \frac{kQ}{a^2}}$$

(*d*) Evaluate $E_z(a)$:

$$E_z(a) = \frac{kQa}{\left[a^2 + a^2\right]^{3/2}} = \boxed{0.354 \frac{kQ}{a^2}}$$

(*e*) Evaluate $E_z(2a)$:

$$E_z(2a) = \frac{2kQa}{\left[(2a)^2 + a^2\right]^{3/2}} = \boxed{0.179 \frac{kQ}{a^2}}$$

(*f*) The field along the *x* axis is plotted below. The *z* coordinates are in units of *z/a* and *E* is in units of kQ/a^2.

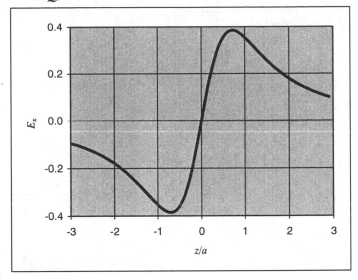

25 •• Calculate the electric field a distance *z* from a uniformly charged infinite flat non-conducting sheet by modeling the sheet as a continuum of infinite circular rings of charge.

Picture the Problem The field at a point on the axis of a uniformly charged ring lies along the axis and is given by Equation 22-8. The diagram shows one ring of the continuum of circular rings of charge. The radius of the ring is *a* and the distance from its center to the field point *P* is *x*. The ring has a uniformly distributed charge *Q*. The resultant electric field at *P* is the sum of the fields due to the continuum of circular rings. Note that, by symmetry, the horizontal components of the electric field cancel.

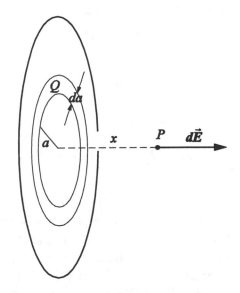

Express the field of a single uniformly charged ring with charge *Q* and radius *a* on the axis of the ring at a distance *x* away from the plane of the ring:

$$\vec{E} = E_x\hat{i} \text{ , where } E_x = \frac{kQx}{\left(x^2 + a^2\right)^{3/2}}$$

Substitute *dq* for *Q* and dE_x for E_x to obtain:

$$dE_x = \frac{kxdq}{\left(x^2 + a^2\right)^{3/2}}$$

The resultant electric field at P is the sum of the fields due to all the circular rings. Integrate both sides to calculate the resultant field for the entire plane. The field point remains fixed, so x is constant:

$$E = \int \frac{kx\,dq}{\left(x^2 + a^2\right)^{3/2}} = kx \int \frac{dq}{\left(x^2 + a^2\right)^{3/2}}$$

To evaluate this integral we change integration variables from q to a. The charge $dq = \sigma\,dA$ where $dA = 2\pi a\,da$ is the area of a ring of radius a and width da:

$$dq = 2\pi\sigma a\,da$$

so

$$E = kx \int_0^\infty \frac{2\pi\sigma a\,da}{\left(x^2 + a^2\right)^{3/2}}$$

$$= 2\pi\sigma kx \int_0^\infty \frac{a\,da}{\left(x^2 + a^2\right)^{3/2}}$$

To integrate this expression, let $u = \sqrt{x^2 + a^2}$. Then:

$$du = \tfrac{1}{2} \frac{1}{\sqrt{x^2 + a^2}}(2a\,da) = \frac{a}{u}\,da$$

or

$$a\,da = u\,du$$

Noting that when $a = 0$, $u = x$, substitute and simplify to obtain:

$$E = 2\pi\sigma kx \int_x^\infty \frac{u}{u^3}\,du = 2\pi\sigma kx \int_x^\infty u^{-2}\,du$$

Evaluating the integral yields:

$$E = 2\pi\sigma kx \left(-\frac{1}{u} \right)\Bigg|_x^\infty = 2\pi k\sigma = \boxed{\frac{\sigma}{2\epsilon_0}}$$

Gauss's Law

29 •   An electric field is given by $\vec{E} = \text{sign}(x)\cdot(300\ \text{N/C})\hat{i}$, where $\text{sign}(x)$ equals -1 if $x < 0$, 0 if $x = 0$, and $+1$ if $x > 0$. A cylinder of length 20 cm and radius 4.0 cm has its center at the origin and its axis along the x axis such that one end is at $x = +10$ cm and the other is at $x = -10$ cm. (*a*) What is the electric flux through each end? (*b*) What is the electric flux through the curved surface of the cylinder? (*c*) What is the electric flux through the entire closed surface? (*d*) What is the net charge inside the cylinder?

Picture the Problem The field at both circular faces of the cylinder is parallel to the outward vector normal to the surface, so the flux is just EA. There is no flux through the curved surface because the normal to that surface is perpendicular to $\vec{E}$. The net flux through the closed surface is related to the net charge inside by Gauss's law.

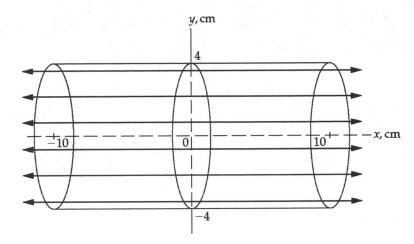

(*a*) Use Gauss's law to calculate the flux through the right circular surface:

$$\phi_{right} = \vec{E}_{right} \cdot \hat{n}_{right} A$$
$$= (300\,\text{N/C})\hat{i} \cdot \hat{i}(\pi)(0.040\,\text{m})^2$$
$$= \boxed{1.5\,\text{N}\cdot\text{m}^2/\text{C}}$$

Apply Gauss's law to the left circular surface:

$$\phi_{left} = \vec{E}_{left} \cdot \hat{n}_{left} A$$
$$= (-300\,\text{N/C})\hat{i} \cdot (-\hat{i})(\pi)(0.040\,\text{m})^2$$
$$= \boxed{1.5\,\text{N}\cdot\text{m}^2/\text{C}}$$

(*b*) Because the field lines are parallel to the curved surface of the cylinder:

$$\phi_{curved} = \boxed{0}$$

(*c*) Express and evaluate the net flux through the entire cylindrical surface:

$$\phi_{net} = \phi_{right} + \phi_{left} + \phi_{curved}$$
$$= 1.5\,\text{N}\cdot\text{m}^2/\text{C} + 1.5\,\text{N}\cdot\text{m}^2/\text{C} + 0$$
$$= \boxed{3.0\,\text{N}\cdot\text{m}^2/\text{C}}$$

(*d*) Apply Gauss's law to obtain:

$$\phi_{net} = 4\pi k Q_{inside} \Rightarrow Q_{inside} = \frac{\phi_{net}}{4\pi k}$$

Substitute numerical values and evaluate Q_{inside}:

$$Q_{inside} = \frac{3.0\,\text{N}\cdot\text{m}^2/\text{C}}{4\pi(8.988\times10^9\,\text{N}\cdot\text{m}^2/\text{C}^2)}$$
$$= \boxed{2.7\times10^{-11}\,\text{C}}$$

33 •   A single point charge is placed at the center of an imaginary cube that has 20-cm-long edges. The electric flux out of one of the cube's sides is $-1.50\,\text{kN}\cdot\text{m}^2/\text{C}$. How much charge is at the center?

Picture the Problem The net flux through the cube is given by $\phi_{net} = Q_{inside}/\epsilon_0$, where Q_{inside} is the charge at the center of the cube.

The flux through one side of the cube is one-sixth of the total flux through the cube:

$$\phi_{1\,faces} = \tfrac{1}{6}\phi_{net} = \frac{Q_{inside}}{6\,\epsilon_0}$$

Solving for Q_{inside} yields:

$$Q_{inside} = 6\,\epsilon_0\,\phi_{2\,faces}$$

Substitute numerical values and evaluate Q_{inside}:

$$Q_{inside} = 6\left(8.854\times10^{-12}\,\frac{C^2}{N\cdot m^2}\right)\left(-1.50\frac{kN\cdot m^2}{C^2}\right) = \boxed{-79.7\,nC}$$

Gauss's Law Applications in Spherical Symmetry Situations

39 ••   A non-conducting sphere of radius 6.00 cm has a uniform volume charge density of 450 nC/m^3. (*a*) What is the total charge on the sphere? Find the electric field at the following distances from the sphere's center: (*b*) 2.00 cm, (*c*) 5.90 cm, (*d*) 6.10 cm, and (*e*) 10.0 cm.

Picture the Problem We can use the definition of volume charge density and the formula for the volume of a sphere to find the total charge of the sphere. Because the charge is distributed uniformly throughout the sphere, we can choose a spherical Gaussian surface and apply Gauss's law to find the electric field as a function of the distance from the center of the sphere.

(*a*) Using the definition of volume charge density, relate the charge on the sphere to its volume:

$$Q = \rho V = \tfrac{4}{3}\pi\rho r^3$$

Substitute numerical values and evaluate Q:

$$Q = \tfrac{4}{3}\pi\left(450\,nC/m^3\right)\left(0.0600\,m\right)^3$$
$$= 0.4072\,nC = \boxed{0.407\,nC}$$

Apply Gauss's law to a spherical surface of radius $r < R$ that is concentric with the spherical shell to obtain:

$$\oint_S E_n\,dA = \frac{1}{\epsilon_0}Q_{inside} \Rightarrow 4\pi r^2 E_n = \frac{Q_{inside}}{\epsilon_0}$$

Solving for E_n yields:

$$E_n = \frac{Q_{inside}}{4\pi \epsilon_0} \frac{1}{r^2} = \frac{kQ_{inside}}{r^2}$$

Because the charge distribution is uniform, we can find the charge inside the Gaussian surface by using the definition of volume charge density to establish the proportion:

$$\frac{Q}{V} = \frac{Q_{inside}}{V'}$$

where V' is the volume of the Gaussian surface.

Solve for Q_{inside} to obtain:

$$Q_{inside} = Q\frac{V'}{V} = Q\frac{r^3}{R^3}$$

Substitute for Q_{inside} to obtain:

$$E_n(r < R) = \frac{Q_{inside}}{4\pi \epsilon_0} \frac{1}{r^2} = \frac{kQ}{R^3}r$$

(b) Evaluate E_n at $r = 2.00$ cm:

$$E_n(2.00\,cm) = \frac{(8.988\times10^9\ N\cdot m^2/C^2)(0.4072\,nC)}{(0.0600\,m)^3}(0.0200\,m) = \boxed{339\,N/C}$$

(c) Evaluate E_n at $r = 5.90$ cm:

$$E_n(5.90\,cm) = \frac{(8.988\times10^9\ N\cdot m^2/C^2)(0.4072\,nC)}{(0.0600\,m)^3}(0.0590\,m) = \boxed{1.00\,kN/C}$$

Apply Gauss's law to the Gaussian surface with $r > R$:

$$4\pi r^2 E_n = \frac{Q_{inside}}{\epsilon_0} \Rightarrow E_n = \frac{kQ_{inside}}{r^2} = \frac{kQ}{r^2}$$

(d) Evaluate E_n at $r = 6.10$ cm:

$$E_n(6.10\,cm) = \frac{(8.988\times10^9\ N\cdot m^2/C^2)(0.4072\,nC)}{(0.0610\,m)^2} = \boxed{983\,N/C}$$

(e) Evaluate E_n at $r = 10.0$ cm:

$$E_n(10.0\,cm) = \frac{(8.988\times10^9\ N\cdot m^2/C^2)(0.4072\,nC)}{(0.100\,m)^2} = \boxed{366\,N/C}$$

43 ••   A sphere of radius R has volume charge density $\rho = B/r$ for $r < R$, where B is a constant and $\rho = 0$ for $r > R$. (a) Find the total charge on the sphere. (b) Find the expressions for the electric field inside and outside the charge distribution (c) Sketch the magnitude of the electric field as a function of the distance r from the sphere's center.

Picture the Problem We can find the total charge on the sphere by expressing the charge dq in a spherical shell and integrating this expression between $r = 0$ and $r = R$. By symmetry, the electric fields must be radial. To find E_r inside the charged sphere we choose a spherical Gaussian surface of radius $r < R$. To find E_r outside the charged sphere we choose a spherical Gaussian surface of radius $r > R$. On each of these surfaces, E_r is constant. Gauss's law then relates E_r to the total charge inside the surface.

(a) Express the charge dq in a shell of thickness dr and volume $4\pi r^2\, dr$:

$$dq = 4\pi r^2 \rho dr = 4\pi r^2 \frac{B}{r} dr$$
$$= 4\pi B r dr$$

Integrate this expression from $r = 0$ to R to find the total charge on the sphere:

$$Q = 4\pi B \int_0^R r dr = \left[2\pi B r^2\right]_0^R$$
$$= \boxed{2\pi B R^2}$$

(b) Apply Gauss's law to a spherical surface of radius $r > R$ that is concentric with the nonconducting sphere to obtain:

$$\oint_S E_r dA = \frac{1}{\epsilon_0} Q_{\text{inside}} \text{ or } 4\pi r^2 E_r = \frac{Q_{\text{inside}}}{\epsilon_0}$$

Solving for E_r yields:

$$E_r(r > R) = \frac{Q_{\text{inside}}}{4\pi \epsilon_0} \frac{1}{r^2} = \frac{kQ_{\text{inside}}}{r^2}$$
$$= \frac{k2\pi B R^2}{r^2} = \boxed{\frac{BR^2}{2\epsilon_0 r^2}}$$

Apply Gauss's law to a spherical surface of radius $r < R$ that is concentric with the nonconducting sphere to obtain:

$$\oint_S E_r dA = \frac{1}{\epsilon_0} Q_{\text{inside}} \Rightarrow 4\pi r^2 E_r = \frac{Q_{\text{inside}}}{\epsilon_0}$$

Solving for E_r yields:

$$E_r(r < R) = \frac{Q_{inside}}{4\pi r^2 \epsilon_0} = \frac{2\pi B r^2}{4\pi r^2 \epsilon_0}$$

$$= \boxed{\frac{B}{2\epsilon_0}}$$

(c) The following graph of E_r versus r/R, with E_r in units of $B/(2\epsilon_0)$, was plotted using a spreadsheet program.

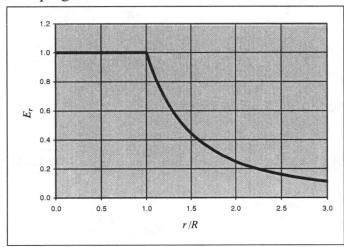

Remarks: Note that our results for (a) and (b) agree at $r = R$.

Gauss's Law Applications in Cylindrical Symmetry Situations

51 •• A solid cylinder of length 200 m and radius 6.00 cm has a uniform volume charge density of 300 nC/m³. (a) What is the total charge of the cylinder? Use the formulas given in Problem 50 to calculate the electric field at a point equidistant from the ends at the following radial distances from the cylindrical axis: (b) 2.00 cm, (c) 5.90 cm, (d) 6.10 cm, and (e) 10.0 cm.

Picture the Problem We can use the definition of volume charge density to find the total charge on the cylinder. From symmetry, the electric field tangent to the surface of the cylinder must vanish. We can construct a Gaussian surface in the shape of a cylinder of radius r and length L and apply Gauss's law to find the electric field as a function of the distance from the centerline of the uniformly charged cylinder.

(a) Use the definition of volume charge density to express the total charge of the cylinder:

$$Q_{tot} = \rho V = \rho(\pi R^2 L)$$

Substitute numerical values to obtain:

$$Q_{tot} = \pi\left(300\,\text{nC/m}^3\right)\left(0.0600\,\text{m}\right)^2\left(200\,\text{m}\right)$$

$$= \boxed{679\,\text{nC}}$$

(b) From Problem 50, for $r < R$, we have:

$$E(r) = \frac{\rho}{2\,\epsilon_0}\,r$$

For $r = 2.00$ cm:

$$E(2.00\,\text{cm}) = \frac{\left(300\,\text{nC/m}^3\right)\left(0.0200\,\text{m}\right)}{2\left(8.854\times10^{-12}\,\text{C}^2/\text{N}\cdot\text{m}^2\right)} = \boxed{339\,\text{N/C}}$$

(c) For $r = 5.90$ cm:

$$E(5.90\,\text{cm}) = \frac{\left(300\,\text{nC/m}^3\right)\left(0.0590\,\text{m}\right)}{2\left(8.854\times10^{-12}\,\text{C}^2/\text{N}\cdot\text{m}^2\right)} = \boxed{1.00\,\text{kN/C}}$$

From Problem 50, for $r > R$, we have:

$$E(r) = \frac{\rho R^2}{2\,\epsilon_0\,r}$$

(d) For $r = 6.10$ cm:

$$E(6.10\,\text{cm}) = \frac{\left(300\,\text{nC/m}^3\right)\left(0.0600\,\text{m}\right)^2}{2\left(8.854\times10^{-12}\,\text{C}^2/\text{N}\cdot\text{m}^2\right)\left(0.0610\,\text{m}\right)} = \boxed{1.00\,\text{kN/C}}$$

(e) For $r = 10.0$ cm:

$$E(10.0\,\text{cm}) = \frac{\left(300\,\text{nC/m}^3\right)\left(0.0600\,\text{m}\right)^2}{2\left(8.854\times10^{-12}\,\text{C}^2/\text{N}\cdot\text{m}^2\right)\left(0.100\,\text{m}\right)} = \boxed{610\,\text{N/C}}$$

55 •• An infinitely long non-conducting solid cylinder of radius a has a non-uniform volume charge density. This density varies with R, the perpendicular distance from its axis, according to $\rho(R) = bR^2$, where b is a constant. (a) Show that the linear charge density of the cylinder is given by $\lambda = \pi b a^4/2$. (b) Find expressions for the electric field for $R < a$ and $R > a$.

Picture the Problem From symmetry; the field tangent to the surface of the cylinder must vanish. We can construct a Gaussian surface in the shape of a cylinder of radius r and length L and apply Gauss's law to find the electric field as a function of the distance from the centerline of the infinitely long nonconducting cylinder.

(a) Apply Gauss's law to a cylindrical surface of radius r and length L that is concentric with the infinitely long nonconducting cylinder:

$$\oint_S E_n \, dA = \frac{1}{\epsilon_0} Q_{inside}$$

or

$$2\pi r L E_n = \frac{Q_{inside}}{\epsilon_0} \Rightarrow E_R = \frac{Q_{inside}}{2\pi r L \, \epsilon_0} \quad (1)$$

where we've neglected the end areas because there is no flux through them.

Express dQ_{inside} for $\rho(r) = br^2$:

$$dQ_{inside} = \rho(r)dV = br^2(2\pi rL)dr$$
$$= 2\pi b r^3 L \, dr$$

Integrate dQ_{inside} from $r = 0$ to R to obtain:

$$Q_{inside} = 2\pi bL \int_0^R r^3 dr = 2\pi bL \left[\frac{r^4}{4} \right]_0^R$$
$$= \frac{\pi bL}{2} R^4$$

Divide both sides of this equation by L to obtain an expression for the charge per unit length λ of the cylinder:

$$\lambda = \frac{Q_{inside}}{L} = \boxed{\frac{\pi bR^4}{2}}$$

(b) Substitute for Q_{inside} in equation (1) and simplify to obtain:

$$E_R(r < R) = \frac{\frac{\pi bL}{2} r^4}{2\pi L r \, \epsilon_0} = \boxed{\frac{b}{4\epsilon_0} r^3}$$

For $r > R$:

$$Q_{inside} = \frac{\pi bL}{2} R^4$$

Substitute for Q_{inside} in equation (1) and simplify to obtain:

$$E_R(r > R) = \frac{\frac{\pi bL}{2} R^4}{2\pi r L \, \epsilon_0} = \boxed{\frac{bR^4}{4r \, \epsilon_0}}$$

57 ••• The inner cylinder of Figure 22-42 is made of non-conducting material and has a volume charge distribution given by $\rho(R) = C/R$, where $C = 200$ nC/m². The outer cylinder is metallic, and both cylinders are infinitely long. (a) Find the charge per unit length (that is, the linear charge density) on the inner cylinder. (b) Calculate the electric field for all values of R.

Picture the Problem We can integrate the density function over the radius of the inner cylinder to find the charge on it and then calculate the linear charge density from its definition. To find the electric field for all values of r we can construct a

Gaussian surface in the shape of a cylinder of radius r and length L and apply Gauss's law to each region of the cable to find the electric field as a function of the distance from its centerline.

(a) Find the charge Q_{inner} on the inner cylinder:

$$Q_{inner} = \int_0^R \rho(r)dV = \int_0^R \frac{C}{r}2\pi rLdr$$

$$= 2\pi CL\int_0^R dr = 2\pi CLR$$

Relate this charge to the linear charge density:

$$\lambda_{inner} = \frac{Q_{inner}}{L} = \frac{2\pi CLR}{L} = 2\pi CR$$

Substitute numerical values and evaluate λ_{inner}:

$$\lambda_{inner} = 2\pi(200\,nC/m)(0.0150\,m)$$

$$= \boxed{18.8\,nC/m}$$

(b) Apply Gauss's law to a cylindrical surface of radius r and length L that is concentric with the infinitely long nonconducting cylinder:

$$\oint_S E_n dA = \frac{1}{\epsilon_0}Q_{inside}$$

or

$$2\pi rLE_n = \frac{Q_{inside}}{\epsilon_0} \Rightarrow E_R = \frac{Q_{inside}}{2\pi rL\,\epsilon_0}$$

where we've neglected the end areas because there is no flux through them.

Substitute to obtain, for $r < 1.50$ cm:

$$E_R(r<1.50\,cm) = \frac{2\pi CLr}{2\pi\,\epsilon_0\,Lr} = \frac{C}{\epsilon_0}$$

Substitute numerical values and evaluate $E_R(r < 1.50$ cm):

$$E_R(r<1.50\,cm) = \frac{200\,nC/m^2}{8.854\times10^{-12}\,C^2/N\cdot m^2}$$

$$= \boxed{22.6\,kN/C}$$

Express Q_{inside} for 1.50 cm $< r <$ 4.50 cm:

$$Q_{inside} = 2\pi CLR$$

Substitute to obtain, for 1.50 cm $< r <$ 4.50 cm:

$$E_R(1.50\,cm < r < 4.50\,cm) = \frac{2C\pi RL}{2\pi\,\epsilon_0\,rL}$$

$$= \frac{CR}{\epsilon_0\,r}$$

where $R = 1.50$ cm.

Substitute numerical values and evaluate $E_n(1.50 \text{ cm} < r < 4.50 \text{ cm})$:

$$E_R(1.50 \text{cm} < r < 4.50 \text{cm}) = \frac{(200 \text{ nC/m}^2)(0.0150 \text{ m})}{(8.854 \times 10^{-12} \text{ C}^2/\text{N} \cdot \text{m}^2)r} = \boxed{\frac{339 \text{ N} \cdot \text{m/C}}{r}}$$

Because the outer cylindrical shell is a conductor:

$$E_R(4.50 \text{cm} < r < 6.50 \text{cm}) = \boxed{0}$$

For $r > 6.50$ cm, $Q_{inside} = 2\pi CLR$ and:

$$E_R(r > 6.50 \text{cm}) = \boxed{\frac{339 \text{ N} \cdot \text{m/C}}{r}}$$

Electric Charge and Field at Conductor Surfaces

63 ••   A positive point charge of 2.5 μC is at the center of a conducting spherical shell that has a net charge of zero, an inner radius equal to 60 cm, and an outer radius equal to 90 cm. (*a*) Find the charge densities on the inner and outer surfaces of the shell and the total charge on each surface. (*b*) Find the electric field everywhere. (*c*) Repeat Part (*a*) and Part (*b*) with a net charge of +3.5 μC placed on the shell.

Picture the Problem Let the inner and outer radii of the uncharged spherical conducting shell be R_1 and R_2 and q represent the positive point charge at the center of the shell. The positive point charge at the center will induce a negative charge on the inner surface of the shell and, because the shell is uncharged, an equal positive charge will be induced on its outer surface. To solve Part (*b*), we can construct a Gaussian surface in the shape of a sphere of radius r with the same center as the shell and apply Gauss's law to find the electric field as a function of the distance from this point. In Part (*c*) we can use a similar strategy with the additional charge placed on the shell.

(*a*) Express the charge density on the inner surface:

$$\sigma_{inner} = \frac{q_{inner}}{A}$$

Express the relationship between the positive point charge q and the charge induced on the inner surface q_{inner}:

$$q + q_{inner} = 0 \Rightarrow q_{inner} = -q$$

Substitute for q_{inner} and A to obtain:

$$\sigma_{inner} = \frac{-q}{4\pi R_1^2}$$

Substitute numerical values and evaluate σ_{inner}:

$$\sigma_{inner} = \frac{-2.5\,\mu C}{4\pi(0.60\,m)^2} = \boxed{-0.55\,\mu C/m^2}$$

Express the charge density on the outer surface:

$$\sigma_{outer} = \frac{q_{outer}}{A}$$

Because the spherical shell is uncharged:

$$q_{outer} + q_{inner} = 0$$

Substitute for q_{outer} to obtain:

$$\sigma_{outer} = \frac{-q_{inner}}{4\pi R_2^2}$$

Substitute numerical values and evaluate σ_{outer}:

$$\sigma_{outer} = \frac{2.5\,\mu C}{4\pi(0.90\,m)^2} = \boxed{0.25\,\mu C/m^2}$$

(b) Apply Gauss's law to a spherical surface of radius r that is concentric with the point charge:

$$\oint_S E_n\,dA = \frac{1}{\epsilon_0}Q_{inside} \Rightarrow 4\pi r^2 E_r = \frac{Q_{inside}}{\epsilon_0}$$

Solve for E_r:

$$E_r = \frac{Q_{inside}}{4\pi r^2 \epsilon_0} \qquad (1)$$

For $r < R_1 = 60$ cm, $Q_{inside} = q$. Substitute in equation (1) and evaluate $E_r(r < 60$ cm$)$ to obtain:

$$E_r(r < 60\,cm) = \frac{q}{4\pi r^2 \epsilon_0} = \frac{kq}{r^2} = \frac{(8.988\times10^9\,N\cdot m^2/C^2)(2.5\,\mu C)}{r^2}$$

$$= \boxed{(2.3\times10^4\,N\cdot m^2/C)\frac{1}{r^2}}$$

Because the spherical shell is a conductor, a charge $-q$ will be induced on its inner surface. Hence, for 60 cm $< r <$ 90 m:

$$Q_{inside} = 0$$
and
$$E_r(60\,cm < r < 90\,cm) = \boxed{0}$$

For $r > 90$ cm, the net charge inside the Gaussian surface is q and:

$$E_r(r > 90\,cm) = \frac{kq}{r^2} = \boxed{(2.3\times10^4\,N\cdot m^2/C)\frac{1}{r^2}}$$

(c) Because $E = 0$ in the conductor:

$$q_{inner} = -2.5\,\mu C$$

and

$$\sigma_{inner} = \boxed{-0.55\,\mu C/m^2}\;\text{as before.}$$

Express the relationship between the charges on the inner and outer surfaces of the spherical shell:

$$q_{outer} + q_{inner} = 3.5\,\mu C$$

and

$$q_{outer} = 3.5\,\mu C - q_{inner} = 6.0\,\mu C$$

σ_{outer} is now given by:

$$\sigma_{outer} = \frac{6.0\,\mu C}{4\pi(0.90\,\text{m})^2} = \boxed{0.59\,\mu C/m^2}$$

For $r < R_1 = 60$ cm, $Q_{inside} = q$ and $E_r(r < 60$ cm$)$ is as it was in (a):

$$E_r(r < 60\,\text{cm}) = \boxed{(2.3\times10^4\,\text{N}\cdot\text{m}^2/\text{C})\frac{1}{r^2}}$$

Because the spherical shell is a conductor, a charge $-q$ will be induced on its inner surface. Hence, for 60 cm $< r <$ 90 cm:

$$Q_{inside} = 0$$

and

$$E_r(60\,\text{cm} < r < 90\,\text{cm}) = \boxed{0}$$

For $r > 0.90$ m, the net charge inside the Gaussian surface is 6.0 μC and:

$$E_r(r > 90\,\text{cm}) = \frac{kq}{r^2} = (8.988\times10^9\,\text{N}\cdot\text{m}^2/\text{C}^2)(6.0\,\mu C)\frac{1}{r^2} = \boxed{(5.4\times10^4\,\text{N}\cdot\text{m}^2/\text{C})\frac{1}{r^2}}$$

65 ••• A square conducting slab carries a net charge of 80 μC. The dimensions of the slab are 1.0 cm $\times$ 5.0 m $\times$ 5.0 m. To the left of the slab is an infinite non-conducting flat sheet with charge density 2.0 $\mu C/m^2$. The faces of the slab are parallel to the sheet. (a) Find the charge on each face of the slab. (Assume that on each face of the slab the surface charge is uniformly distributed, and that the amount of charge on the edges of the slab is negligible.) (b) Find the electric field just to the left of the slab and just to the right of the slab.

Picture the Problem (*a*) We can use the fact that the net charge on the conducting slab is the sum of the charges Q_{left} and Q_{right} on its left and right surfaces to obtain a linear equation relating these charges. Because the electric field is zero inside the slab, we can obtain a second linear equation in Q_{left} and Q_{right} that we can solve simultaneously with the first equation to find Q_{left} and Q_{right}. (*b*) We can find the electric field on each side of the slab by adding the fields due to the charges on the surfaces of the slab and the field due to the plane of charge.

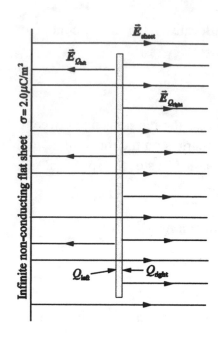

(*a*) The net charge on the conducting slab is the sum of the charges on the surfaces to the left and to the right:

$$Q_{left} + Q_{right} = 80 \ \mu C \qquad (1)$$

Because the electric field is equal to zero inside the slab:

$$\frac{\sigma_{sheet}}{2\,\epsilon_0} + \frac{\sigma_{left}}{2\,\epsilon_0} - \frac{\sigma_{right}}{2\,\epsilon_0} = 0$$

Letting A represent the area of the charged surfaces of the slab and substituting for σ_{left} and σ_{rightt} yields:

$$\frac{\sigma_{sheet}}{2\,\epsilon_0} + \frac{Q_{left}}{2A\,\epsilon_0} - \frac{Q_{right}}{2A\,\epsilon_0} = 0$$

Simplifying to obtain:

$$A\sigma_{sheet} + Q_{left} - Q_{right} = 0$$

Substituting numerical values yields:

$$(5.0 \text{ m})^2 \left(2.0\,\frac{\mu C}{\text{m}^2}\right) + Q_{left} - Q_{right} = 0$$

or

$$Q_{left} - Q_{right} = -50\,\mu C \qquad (2)$$

Solve equations (1) and (2) simultaneously to obtain:

$$Q_{left} = \boxed{15\,\mu C} \text{ and } Q_{right} = \boxed{65\,\mu C}$$

(*b*) Express the total field just to the left of the slab:

$$\vec{E}_{\substack{\text{left of} \\ \text{the slab}}} = \vec{E}_{\text{sheet}} + \vec{E}_{Q_{\text{left}}} + \vec{E}_{Q_{\text{right}}}$$

$$= \frac{\sigma_{\text{sheet}}}{2\epsilon_0}\hat{r} - \frac{\sigma_{\text{left}}}{2\epsilon_0}\hat{r} - \frac{\sigma_{\text{right}}}{2\epsilon_0}\hat{r}$$

where $\hat{r}$ is a unit vector pointing away from the slab.

Substituting for σ_{left} and σ_{right} and simplifying yields:

$$\vec{E}_{\substack{\text{left of} \\ \text{the slab}}} = \frac{\sigma_{\text{sheet}}}{2c_0}\hat{r} - \frac{Q_{\text{left}}}{2A\epsilon_0}\hat{r} - \frac{Q_{\text{right}}}{2A\epsilon_0}\hat{r}$$

$$= \frac{A\sigma_{\text{sheet}} - \left(Q_{\text{left}} + Q_{\text{right}}\right)}{2A\epsilon_0}\hat{r}$$

Substitute numerical values and evaluate $\vec{E}_{\substack{\text{left of} \\ \text{the slab}}}$:

$$\vec{E}_{\substack{\text{left of} \\ \text{the slab}}} = \frac{\left(5.0\,\text{m}\right)^2\left(2.0\,\mu\text{C/m}^2\right) - \left(15\,\mu\text{C} + 65\,\mu\text{C}\right)}{2\left(5.0\,\text{m}\right)^2\left(8.854\times10^{-12}\,\text{C}^2/\text{N}\cdot\text{m}^2\right)}\hat{r} = \boxed{\left(-68\,\text{kN/C}\right)\hat{r}}$$

Express the total field just to the right of the slab:

$$\vec{E}_{\substack{\text{right of} \\ \text{the slab}}} = \vec{E}_{\text{sheet}} + \vec{E}_{Q_{\text{left}}} + \vec{E}_{\substack{\text{right surface} \\ \text{of the slab}}}$$

$$= \frac{\sigma_{\text{sheet}}}{2\epsilon_0}\hat{r} + \frac{\sigma_{\text{left}}}{2\epsilon_0}\hat{r} + \frac{\sigma_{\text{right}}}{2\epsilon_0}\hat{r}$$

Substituting for σ_{left} and σ_{right} and simplifying yields:

$$\vec{E}_{\substack{\text{right of} \\ \text{the slab}}} = \frac{\sigma_{\text{sheet}}}{2\epsilon_0}\hat{r} + \frac{Q_{\text{left}}}{2A\epsilon_0}\hat{r} + \frac{Q_{\text{right}}}{2A\epsilon_0}\hat{r}$$

$$= \frac{A\sigma_{\text{sheet}} + \left(Q_{\text{right}} + Q_{\text{left}}\right)}{2A\epsilon_0}\hat{r}$$

Substitute numerical values and evaluate $\vec{E}_{\substack{\text{right of} \\ \text{the slab}}}$:

$$\vec{E}_{\substack{\text{right of} \\ \text{the slab}}} = \frac{\left(5.0\,\text{m}\right)^2\left(2.0\,\mu\text{C/m}^2\right) + \left(65\,\mu\text{C} + 15\,\mu\text{C}\right)}{2\left(5.0\,\text{m}\right)^2\left(8.854\times10^{-12}\,\text{C}^2/\text{N}\cdot\text{m}^2\right)}\hat{r} = \boxed{\left(0.29\,\text{MN/C}\right)\hat{r}}$$

General Problems

67 •• A large, flat, nonconducting, non-uniformly charged surface lies in the $x = 0$ plane. At the origin, the surface charge density is $+3.10\ \mu\text{C/m}^2$. A small distance away from the surface on the positive x axis, the x component of the electric field is 4.65×10^5 N/C. What is E_x a small distance away from the surface on the negative x axis?

Picture the Problem Because the difference between the field just to the right of the surface $E_{x,pos}$ and the field just to the left of the surface $E_{x,neg}$ is the field due to the nonuniform surface charge, we can express $E_{x,neg}$ as the difference between $E_{x,pos}$ and σ/ϵ_0.

Express the electric field just to the left of the origin in terms of $E_{x,pos}$ and σ/ϵ_0:

$$E_{x,neg} = E_{x,pos} - \frac{\sigma}{\epsilon_0}$$

Substitute numerical values and evaluate $E_{x,neg}$:

$$E_{x,neg} = 4.65 \times 10^5 \text{ N/C} - \frac{3.10 \,\mu\text{C/m}^2}{8.854 \times 10^{-12} \text{ C}^2/\text{N} \cdot \text{m}^2} = \boxed{115 \text{ kN/C}}$$

69 •• A thin, non-conducting, uniformly charged spherical shell of radius R (Figure 22-44a) has a total positive charge of Q. A small circular plug is removed from the surface. (a) What is the magnitude and direction of the electric field at the center of the hole? (b) The plug is now put back in the hole (Figure 22-44b). Using the result of Part (a), find the electric force acting on the plug. (c) Using the magnitude of the force, calculate the "electrostatic pressure" (force/unit area) that tends to expand the sphere.

Picture the Problem If the patch is small enough, the field at the center of the patch comes from two contributions. We can view the field in the hole as the sum of the field from a uniform spherical shell of charge Q plus the field due to a small patch with surface charge density equal but opposite to that of the patch cut out.

(a) Express the magnitude of the electric field at the center of the hole:

$$E = E_{\substack{\text{spherical} \\ \text{shell}}} + E_{\text{hole}}$$

Apply Gauss's law to a spherical gaussian surface just outside the given sphere:

$$E_{\substack{\text{spherical} \\ \text{shell}}} \left(4\pi r^2\right) = \frac{Q_{\text{enclosed}}}{\epsilon_0} = \frac{Q}{\epsilon_0}$$

Solve for $E_{\substack{\text{spherical} \\ \text{shell}}}$ to obtain:

$$E_{\substack{\text{spherical} \\ \text{shell}}} = \frac{Q}{4\pi \epsilon_0 \, r^2}$$

The electric field due to the small hole (small enough so that we can treat it as a plane surface) is:

$$E_{\text{hole}} = \frac{-\sigma}{2\epsilon_0}$$

Substitute and simplify to obtain:

$$E = \frac{Q}{4\pi \epsilon_0 \, r^2} + \frac{-\sigma}{2\epsilon_0}$$

$$= \frac{Q}{4\pi \epsilon_0 \, r^2} - \frac{Q}{2\epsilon_0 \left(4\pi r^2\right)}$$

$$= \boxed{\frac{Q}{8\pi \epsilon_0 \, r^2}} \text{ radially outward}$$

(b) Express the force on the patch:

$$F = qE$$

where q is the charge on the patch.

Assuming that the patch has radius a, express the proportion between its charge and that of the spherical shell:

$$\frac{q}{\pi a^2} = \frac{Q}{4\pi r^2} \text{ or } q = \frac{a^2}{4r^2}Q$$

Substitute for q and E in the expression for F to obtain:

$$F = \left(\frac{a^2}{4r^2}Q\right)\left(\frac{Q}{8\pi \epsilon_0 \, r^2}\right)$$

$$= \boxed{\frac{Q^2 a^2}{32\pi \epsilon_0 \, r^4}} \text{ radially outward}$$

(c) The pressure is the force exerted on the patch divided by the area of the patch:

$$P = \frac{\dfrac{Q^2 a^2}{32\pi \epsilon_0 \, r^4}}{\pi a^2} = \boxed{\frac{Q^2}{32\pi^2 \epsilon_0 \, r^4}}$$

73 •••   A quantum-mechanical treatment of the hydrogen atom shows that the electron in the atom can be treated as a smeared-out distribution of negative charge of the form $\rho(r) = -\rho_0 e^{-2r/a}$. Here r represents the distance from the center of the nucleus and a represents the first *Bohr radius* which has a numerical value of 0.0529 nm. Recall that the nucleus of a hydrogen atom consists of just one proton and treat this proton as a positive point charge.
(a) Calculate ρ_0, using the fact that the atom is neutral. (b) Calculate the electric field at any distance r from the nucleus.

Picture the Problem Because the atom is uncharged, we know that the integral of the electron's charge distribution over all of space must equal its charge q_e. Evaluation of this integral will lead to an expression for ρ_0. In (b) we can express the resultant electric field at any point as the sum of the electric fields due to the proton and the electron cloud.

(a) Because the atom is uncharged, the integral of the electron's charge distribution over all of space must equal its charge e:

$$e = \int_0^\infty \rho(r)dV = \int_0^\infty \rho(r)4\pi r^2 dr$$

Substitute for $\rho(r)$ and simplify to obtain:

$$e = -\int_0^\infty \rho_0 e^{-2r/a} 4\pi r^2 dr$$

$$= -4\pi\rho_0 \int_0^\infty r^2 e^{-2r/a} dr$$

Use integral tables or integration by parts to obtain:

$$\int_0^\infty r^2 e^{-2r/a} dr = \frac{a^3}{4}$$

Substitute for $\int_0^\infty r^2 e^{-2r/a} dr$ to obtain:

$$e = -4\pi\rho_0 \left(\frac{a^3}{4}\right) = -\pi a^3 \rho_0$$

Solving for ρ_0 yields:

$$\boxed{\rho_0 = -\frac{e}{\pi a^3}}$$

(b) The field will be the sum of the field due to the proton and that of the electron charge cloud:

$$E = E_p + E_{cloud}$$

Express the field due to the electron cloud:

$$E_{cloud}(r) = \frac{kQ(r)}{r^2}$$

where $Q(r)$ is the net negative charge enclosed a distance r from the proton.

Substitute for E_p and E_{cloud} to obtain:

$$E(r) = \frac{ke}{r^2} + \frac{kQ(r)}{r^2} \qquad (1)$$

$Q(r)$ is given by:

$$Q(r) = \int_0^r 4\pi r'^2 \rho(r')dr'$$

$$= 4\pi \int_0^r r'^2 \rho_0 e^{-2r'/a} dr'$$

From Part (a), $\rho_0 = \frac{-e}{\pi a^3}$:

$$Q(r) = 4\pi\left(\frac{-e}{\pi a^3}\right)\int_0^r r'^2 e^{-2r'/a} dr'$$

$$= \frac{-4e}{a^3}\int_0^r r'^2 e^{-2r'/a} dr'$$

From a table of integrals:

$$\int_0^r x^2 e^{-2x/a}\,dx = \tfrac{1}{4}e^{-2r/a}\,a\left[\left(e^{-2r/a}-1\right)a^2 - 2ar - 2r^2\right]$$

$$= \tfrac{1}{4}e^{-2r/a}\,a^3\left[\left(e^{-2r/a}-1\right) - 2\frac{r}{a} - 2\frac{r^2}{a^2}\right]$$

$$= \frac{a^3}{4}\left[\left(1-e^{-2r/a}\right) - 2e^{-2r/a}\left(\frac{r}{a}+\frac{r^2}{a^2}\right)\right]$$

Substituting for $\int_0^r r'^2 e^{-2r'/a}\,dr'$ in the expression for $Q(r)$ and simplifying yields:

$$Q(r) = \frac{-e}{4}\left[\left(1-e^{-2r/a}\right) - 2e^{-2r/a}\left(\frac{r}{a}+\frac{r^2}{a^2}\right)\right]$$

Substitute for $Q(r)$ in equation (1) and simplify to obtain:

$$E(r) = \frac{ke}{r^2} - \frac{ke}{4r^2}\left[\left(1-e^{-2r/a}\right) - 2e^{-2r/a}\left(\frac{r}{a}+\frac{r^2}{a^2}\right)\right]$$

$$= \boxed{\frac{ke}{r^2}\left(1 - \frac{1}{4}\left[\left(1-e^{-2r/a}\right) - 2e^{-2r/a}\left(\frac{r}{a}+\frac{r^2}{a^2}\right)\right]\right)}$$

79 •• A uniformly charged, infinitely long line of negative charge has a linear charge density of $-\lambda$ and is located on the z axis. A small positively charged particle that has a mass m and a charge q is in a circular orbit of radius R in the xy plane centered on the line of charge. (a) Derive an expression for the speed of the particle. (b) Obtain an expression for the period of the particle's orbit.

Picture the Problem (a) We can apply Newton's 2nd law to the particle to express its speed as a function of its mass m, charge q, and the radius of its path R, and the strength of the electric field due to the infinite line charge E. (b) The period of the particle's motion is the ratio of the circumference of the circle in which it travels divided by its orbital speed.

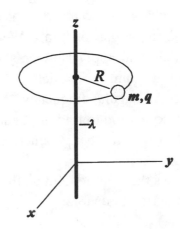

(a) Apply Newton's 2nd law to the particle to obtain:

$$\sum F_{radial} = qE = m\frac{v^2}{R}$$

where the inward direction is positive.

Solving for v yields:

$$v = \sqrt{\frac{qRE}{m}}$$

The strength of the electric field at a distance R from the infinite line charge is given by:

$$E = \frac{2k\lambda}{R}$$

Substitute for E and simplify to obtain:

$$\boxed{v = \sqrt{\frac{2kq\lambda}{m}}}$$

(b) The speed of the particle is equal to the circumference of its orbit divided by its period:

$$v = \frac{2\pi R}{T} \Rightarrow T = \frac{2\pi R}{v}$$

Substitute for v and simplify to obtain:

$$\boxed{T = \pi R \sqrt{\frac{2m}{kq\lambda}}}$$

81 •• The charges Q and q of Problem 80 are +5.00 μC and –5.00 μC, respectively, and the radius of the ring is 8.00 cm. When the particle is given a small displacement in the x direction, it oscillates about its equilibrium position at a frequency of 3.34 Hz. (a) What is the particle's mass? (b) What is the frequency if the radius of the ring is doubled to 16.0 cm and all other parameters remain unchanged?

Picture the Problem Starting with the equation for the electric field on the axis of a ring charge, we can factor the denominator of the expression to show that, for $x \ll a$, E_x is proportional to x. We can use $F_x = qE_x$ to express the force acting on the particle and apply Newton's 2nd law to show that, for small displacements from equilibrium, the particle will execute simple harmonic motion. Finally, we can find the angular frequency of the motion from the differential equation and use this expression to find the frequency of the motion when the radius of the ring is doubled and all other parameters remain unchanged.

(a) Express the electric field on the axis of the ring of charge:

$$E_x = \frac{kQx}{\left(x^2 + a^2\right)^{3/2}}$$

Factor a^2 from the denominator of E_x to obtain:

$$E_x = \frac{kQx}{\left[a^2\left(1+\dfrac{x^2}{a^2}\right)\right]^{3/2}}$$

$$= \frac{kQx}{a^3\left(1+\dfrac{x^2}{a^2}\right)^{3/2}} \approx \frac{kQ}{a^3}x$$

provided $x << a$.

Express the force acting on the particle as a function of its charge and the electric field:

$$F_x = qE_x = \frac{kqQ}{a^3}x$$

Because the negatively charged particle experiences a linear restoring force, we know that its motion will be simple harmonic. Apply Newton's 2nd law to the negatively charged particle to obtain:

$$m\frac{d^2x}{dt^2} = -\frac{kqQ}{a^3}x$$

or

$$\frac{d^2x}{dt^2} + \frac{kqQ}{ma^3}x = 0$$

the differential equation of simple harmonic motion.

The angular frequency of the simple harmonic motion of the particle is given by:

$$\omega = \sqrt{\frac{kqQ}{ma^3}} \qquad (1)$$

Solving for m yields:

$$m = \frac{kqQ}{\omega^2 a^3} = \frac{kqQ}{4\pi^2 f^2 a^3}$$

Substitute numerical values and evaluate m:

$$m = \frac{\left(8.988\times10^9\ \dfrac{\text{N}\cdot\text{m}^2}{\text{C}^2}\right)|(-5.00\,\mu\text{C}|)(5.00\,\mu\text{C})}{4\pi^2(3.34\,\text{s}^{-1})^2(8.00\,\text{cm})^3} = \boxed{0.997\ \text{kg}}$$

(b) Express the angular frequency of the motion if the radius of the ring is doubled:

$$\omega' = \sqrt{\frac{kqQ}{m(2a)^3}} \qquad (2)$$

Divide equation (2) by equation (1) to obtain:

$$\frac{\omega'}{\omega} = \frac{2\pi f'}{2\pi f} = \frac{\sqrt{\dfrac{kqQ}{m(2a)^3}}}{\sqrt{\dfrac{kqQ}{ma^3}}} = \frac{1}{\sqrt{8}}$$

Solve for f' to obtain:

$$f' = \frac{f}{\sqrt{8}} = \frac{3.34\,\text{Hz}}{\sqrt{8}} = \boxed{1.18\,\text{Hz}}$$

87 •••   Consider a simple but surprisingly accurate model for the hydrogen molecule: two positive point charges, each having charge $+e$, are placed inside a uniformly charged sphere of radius R, which has a charge equal to $-2e$. The two point charges are placed symmetrically, equidistant from the center of the sphere (Figure 22-48). Find the distance from the center, a, where the net force on either point charge is zero.

Picture the Problem We can find the distance from the center where the net force on either charge is zero by setting the sum of the forces acting on either point charge equal to zero. Each point charge experiences two forces; one a Coulomb force of repulsion due to the other point charge, and the second due to that fraction of the sphere's charge that is between the point charge and the center of the sphere that creates an electric field at the location of the point charge.

Apply $\sum F = 0$ to either of the point charges:

$$F_{\text{Coulomb}} - F_{\text{field}} = 0 \qquad (1)$$

Express the Coulomb force on the proton:

$$F_{\text{Coulomb}} = \frac{ke^2}{(2a)^2} = \frac{ke^2}{4a^2}$$

The force exerted by the field E is:

$$F_{\text{field}} = eE$$

Apply Gauss's law to a spherical surface of radius a centered at the origin:

$$E(4\pi a^2) = \frac{Q_{\text{enclosed}}}{\epsilon_0}$$

Relate the charge density of the electron sphere to Q_{enclosed}:

$$\frac{2e}{\frac{4}{3}\pi R^3} = \frac{Q_{\text{enclosed}}}{\frac{4}{3}\pi a^3} \Rightarrow Q_{\text{enclosed}} = \frac{2ea^3}{R^3}$$

Substitute for Q_{enclosed}:

$$E(4\pi a^2) = \frac{2ea^3}{\epsilon_0 R^3}$$

Solve for E to obtain:

$$E = \frac{ea}{2\pi \epsilon_0 R^3} \Rightarrow F_{\text{field}} = \frac{e^2 a}{2\pi \epsilon_0 R^3}$$

Substitute for F_{Coulomb} and F_{field} in equation (1):

$$\frac{ke^2}{4a^2} - \frac{e^2 a}{2\pi \,\epsilon_0\, R^3} = 0$$

or

$$\frac{ke^2}{4a^2} - \frac{2ke^2 a}{R^3} = 0 \Rightarrow a = \sqrt[3]{\frac{1}{8}} R = \boxed{\tfrac{1}{2} R}$$

Chapter 23
Electrical Potential

Conceptual Problems

1 • A proton is moved to the left in a uniform electric field that points to the right. Is the proton moving in the direction of increasing or decreasing electric potential? Is the electrostatic potential energy of the proton increasing or decreasing?

Determine the Concept The proton is moving to a region of higher potential. The proton's electrostatic potential energy is increasing.

5 •• Figure 23-29 shows a point particle that has a positive charge +Q and a metal sphere that has a charge –Q. Sketch the electric field lines and equipotential surfaces for this system of charges.

Picture the Problem The electric field lines, shown as solid lines, and the equipotential surfaces (intersecting the plane of the paper), shown as dashed lines, are sketched in the adjacent figure. The point charge +Q is the point at the right, and the metal sphere with charge –Q is at the left. Near the two charges the equipotential surfaces are spheres, and the field lines are normal to the metal sphere at the sphere's surface.

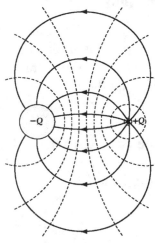

11 •• The electric potential is the same everywhere on the surface of a conductor. Does this mean that the surface charge density is also the same everywhere on the surface? Explain your answer.

Determine the Concept No. The local surface charge density is proportional to the normal component of the electric field, not the potential on the surface.

Estimation and Approximation Problems

13 • Estimate maximum the potential difference between a thundercloud and Earth, given that the electrical breakdown of air occurs at fields of roughly 3.0×10^6 V/m.

Picture the Problem The field of a thundercloud must be of order 3.0×10^6 V/m just before a lightning strike.

Express the potential difference between the cloud and the earth as a function of their separation d and electric field E between them:	$V = Ed$

Assuming that the thundercloud is at a distance of about 1 km above the surface of the earth, the potential difference is approximately:	$V = (3.0 \times 10^6 \text{ V/m})(10^3 \text{ m})$ $= \boxed{3.0 \times 10^9 \text{ V}}$

Note that this is an upper bound, as there will be localized charge distributions on the thundercloud which raise the local electric field above the average value.

Electrostatic Potential Difference, Electrostatic Energy and Electric Field

23 ••   Protons are released from rest in a Van de Graaff accelerator system. The protons initially are located where the electrical potential has a value of 5.00 MV and then they travel through a vacuum to a region where the potential is zero. (*a*) Find the final speed of these protons. (*b*) Find the accelerating electric-field strength if the potential changed *uniformly* over a distance of 2.00 m.

Picture the Problem We can find the final speeds of the protons from the potential difference through which they are accelerated and use $E = \Delta V / \Delta x$ to find the accelerating electric field.

(*a*) Apply the work-kinetic energy theorem to the accelerated protons:	$W = \Delta K = K_f \Rightarrow e\Delta V = \frac{1}{2}mv^2$

Solve for v to obtain:	$v = \sqrt{\dfrac{2e\Delta V}{m}}$

Substitute numerical values and evaluate v:	$v = \sqrt{\dfrac{2(1.602 \times 10^{-19} \text{ C})(5.00\,\text{MV})}{1.673 \times 10^{-27} \text{ kg}}}$ $= \boxed{3.09 \times 10^7 \text{ m/s}}$

(*b*) Assuming the same potential change occurred *uniformly* over the distance of 2.00 m, we can use the relationship between E, ΔV, and Δx express and evaluate E:	$E = \dfrac{\Delta V}{\Delta x} = \dfrac{5.00\,\text{MV}}{2.00\,\text{m}} = \boxed{2.50\,\text{MV/m}}$

Potential Due to a System of Point Charges

27 •   Three point charges are fixed at locations on the x-axis: q_1 is at $x = 0.00$ m, q_2 is at $x = 3.00$ m, and q_3 is at $x = 6.00$ m. Find the electric potential at the point on the y axis at $y = 3.00$ m if (a) $q_1 = q_2 = q_3 = +2.00$ μC, (b) $q_1 = q_2 = +2.00$ μC and $q_3 = -2.00$ μC, and (c) $q_1 = q_3 = +2.00$ μC and $q_2 = -2.00$ μC. (Assume the potential is zero very far from all charges.)

Picture the Problem The potential at the point whose coordinates are (0, 3.00 m) is the algebraic sum of the potentials due to the charges at the three locations given.

Express the potential at the point whose coordinates are (0, 3.00 m):

$$V = k\sum_{i=1}^{3}\frac{q_i}{r_i} = k\left(\frac{q_1}{r_1} + \frac{q_2}{r_2} + \frac{q_3}{r_3}\right)$$

(a) For $q_1 = q_2 = q_3 = 2.00$ μC:

$$V = \left(8.988\times10^9\ \text{N}\cdot\text{m}^2/\text{C}^2\right)\left(2.00\,\mu\text{C}\right)\left(\frac{1}{3.00\,\text{m}} + \frac{1}{3.00\sqrt{2}\,\text{m}} + \frac{1}{3.00\sqrt{5}\,\text{m}}\right)$$

$$= \boxed{12.9\,\text{kV}}$$

(b) For $q_1 = q_2 = 2.00$ μC and $q_3 = -2.00$ μC:

$$V = \left(8.988\times10^9\ \text{N}\cdot\text{m}^2/\text{C}^2\right)\left(2.00\,\mu\text{C}\right)\left(\frac{1}{3.00\,\text{m}} + \frac{1}{3.00\sqrt{2}\,\text{m}} - \frac{1}{3.00\sqrt{5}\,\text{m}}\right)$$

$$= \boxed{7.55\,\text{kV}}$$

(c) For $q_1 = q_3 = 2.00$ μC and $q_2 = -2.00$ μC:

$$V = \left(8.988\times10^9\ \text{N}\cdot\text{m}^2/\text{C}^2\right)\left(2.00\,\mu\text{C}\right)\left(\frac{1}{3.00\,\text{m}} - \frac{1}{3.00\sqrt{2}\,\text{m}} + \frac{1}{3.00\sqrt{5}\,\text{m}}\right)$$

$$= \boxed{4.43\,\text{kV}}$$

31 ••   Two identical positively charged point particles are fixed on the x-axis at $x = +a$ and $x = -a$. (a) Write an expression for the electric potential $V(x)$ as a function of x for all points on the x-axis. (b) Sketch $V(x)$ versus x for all points on the x axis.

Picture the Problem For the two charges, $r = |x - a|$ and $|x + a|$ respectively and the electric potential at x is the algebraic sum of the potentials at that point due to the charges at $x = +a$ and $x = -a$.

(*a*) Express $V(x)$ as the sum of the potentials due to the charges at $x = +a$ and $x = -a$:

$$V = \left| kq\left(\frac{1}{|x - a|} + \frac{1}{|x + a|}\right) \right|$$

(*b*) The following graph of V as a function of x/a was plotted using a spreadsheet program:

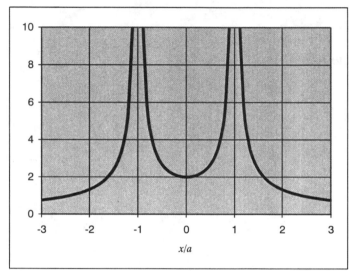

33 •••   A dipole consists of equal but opposite point charges $+q$ and $-q$. It is located so that its center is at the origin, and its axis is aligned with the z-axis (Figure 23-32) The distance between the charges is L. Let $\vec{r}$ be the vector from the origin to an arbitrary field point and θ be the angle that $\vec{r}$ makes with the $+z$ direction. (*a*) Show that at large distances from the dipole (that is for $r \gg L$), the dipole's electric potential is given by $V(r, \theta) \approx k\vec{p} \cdot \hat{r}/r^2 = kp\cos\theta/r^2$, where $\vec{p}$ is the dipole moment of the dipole and θ is the angle between $\vec{r}$ and $\vec{p}$. (*b*) At what points in the region $r \gg L$, other than at infinity, is the electric potential zero?

Picture the Problem The potential at the arbitrary field point is the sum of the potentials due to the equal but opposite point charges.

(*a*) Express the potential at the arbitrary field point at a large distance from the dipole:

$$V = V_+ + V_- = \frac{kq}{r_+} + \frac{k(-q)}{r_-}$$

$$= kq\left(\frac{1}{r_+} - \frac{1}{r_-}\right) = kq\left(\frac{r_- - r_+}{r_+ r_-}\right)$$

Referring to the figure, note that, for the far field ($r \gg L$):

$$r_- - r_+ \approx L\cos\theta \text{ and } r_+ \approx r_- \approx r$$

Substituting and simplifying yields:

$$V(r,\theta) = kq\left(\frac{L\cos\theta}{r^2}\right) = \frac{kqL\cos\theta}{r^2}$$

or, because $p = qL$,

$$V(r,\theta) = \frac{kp\cos\theta}{r^2}$$

Finally, because $\vec{p}\cdot\hat{r} = p\cos\theta$:

$$V(r,\theta) = \boxed{\frac{k\vec{p}\cdot\hat{r}}{r^2}}$$

(b) $V(r,\theta) = 0$ where $\cos\theta = 0$:

$\theta = \cos^{-1}0 = 90° \Rightarrow V = 0$ at points on the z axis. Note that these locations are equidistant from the two oppositely-charged ends of the dipole.

Calculations of V for Continuous Charge Distributions

41 • An infinite line charge of linear charge density $+1.50~\mu C/m$ lies on the z-axis. Find the electric potential at distances from the line charge of (a) 2.00 m, (b) 4.00 m, and (c) 12.0 m. Assume that we choose $V = 0$ at a distance of 2.50 m from the line of charge.

Picture the Problem We can use the expression for the potential due to a line charge $V(r) = -2k\lambda\ln\left(\dfrac{r}{a}\right)$, where $V = 0$ at some distance $r = a$, to find the potential at these distances from the line.

Express the potential due to a line charge as a function of the distance from the line:

$$V(r) = -2k\lambda\ln\left(\frac{r}{a}\right)$$

Because $V = 0$ at $r = 2.50$ m:

$$0 = -2k\lambda\ln\left(\frac{2.50\,\text{m}}{a}\right)$$

$$\Rightarrow 0 = \ln\left(\frac{2.50\,\text{m}}{a}\right)$$

and

$$\frac{2.50\,\text{m}}{a} = \ln^{-1}(0) = 1 \Rightarrow a = 2.50 \text{ m}$$

Thus we have $a = 2.50$ m and:

$$V(r) = -2 \left(8.988 \times 10^9 \, \frac{\text{N} \cdot \text{m}^2}{\text{C}^2} \right) \left(1.5 \frac{\mu\text{C}}{\text{m}} \right) \ln\left(\frac{r}{2.50\,\text{m}} \right)$$

$$= -\left(2.696 \times 10^4 \, \text{N} \cdot \text{m/C} \right) \ln\left(\frac{r}{2.50\,\text{m}} \right)$$

(a) Evaluate $V(2.00$ m):

$$V(2.00\,\text{m}) = -\left(2.696 \times 10^4 \, \frac{\text{N} \cdot \text{m}}{\text{C}} \right)$$

$$\times \ln\left(\frac{2.00\,\text{m}}{2.50\,\text{m}} \right)$$

$$= \boxed{6.02\,\text{kV}}$$

(b) Evaluate $V(4.00$ m):

$$V(4.00\,\text{m}) = -\left(2.696 \times 10^4 \, \frac{\text{N} \cdot \text{m}}{\text{C}} \right)$$

$$\times \ln\left(\frac{4.00\,\text{m}}{2.50\,\text{m}} \right)$$

$$= \boxed{-12.7\,\text{kV}}$$

(c) Evaluate $V(12.0$ m):

$$V(12.0\,\text{m}) = -\left(2.696 \times 10^4 \, \frac{\text{N} \cdot \text{m}}{\text{C}} \right)$$

$$\times \ln\left(\frac{12.0\,\text{m}}{2.50\,\text{m}} \right)$$

$$= \boxed{-42.3\,\text{kV}}$$

45 •• Two coaxial conducting cylindrical shells have equal and opposite charges. The inner shell has charge $+q$ and an outer radius a, and the outer shell has charge $-q$ and an inner radius b. The length of each cylindrical shell is L, and L is very long compared with b. Find the potential difference, $V_a - V_b$ between the shells.

Picture the Problem The diagram is a cross-sectional view showing the charges on the inner and outer conducting shells. A portion of the Gaussian surface over which we'll integrate E in order to find V in the region $a < r < b$ is also shown. Once we've determined how E varies with r, we can find $V_a - V_b$ from $V_b - V_a = -\int E_r dr$.

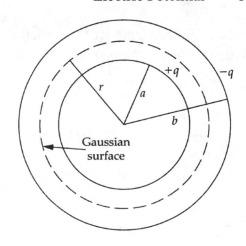

Express the potential difference $V_b - V_a$:

$$V_b - V_a = -\int E_r dr \Rightarrow V_a - V_b = \int E_r dr$$

Apply Gauss's law to cylindrical Gaussian surface of radius r and length L:

$$\oint_S \vec{E} \cdot \hat{n} dA = E_r(2\pi r L) = \frac{q}{\epsilon_0}$$

Solving for E_r yields:

$$E_r = \frac{q}{2\pi \epsilon_0 rL}$$

Substitute for E_r and integrate from $r = a$ to b:

$$V_a - V_b = \frac{q}{2\pi \epsilon_0 L} \int_a^b \frac{dr}{r}$$

$$= \boxed{\frac{2kq}{L} \ln\left(\frac{b}{a}\right)}$$

51 ••   A rod of length L has a total charge Q uniformly distributed along its length. The rod lies along the y-axis with its center at the origin. (*a*) Find an expression for the electric potential as a function of position along the x-axis. (*b*) Show that the result obtained in Part (*a*) reduces to $V = kQ/|x|$ for $|x| \gg L$. Explain why this result is expected.

Picture the Problem Let the charge per unit length be $\lambda = Q/L$ and dy be a line element with charge λdy. We can express the potential dV at any point on the x axis due to the charge element λdy and integrate to find $V(x, 0)$.

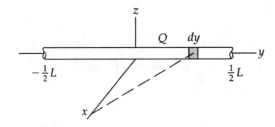

(*a*) Express the element of potential *dV* due to the line element *dy*:

$$dV = \frac{k\lambda}{r}\,dy$$

where $r = \sqrt{x^2 + y^2}$ and $\lambda = \dfrac{Q}{L}$.

Substituting for *r* and λ yields:

$$dV = \frac{kQ}{L}\frac{dy}{\sqrt{x^2+y^2}}$$

Use a table of integrals to integrate *dV* from $y = -L/2$ to $y = L/2$:

$$V(x,0) = \frac{kQ}{L}\int_{-L/2}^{L/2}\frac{dy}{\sqrt{x^2+y^2}}$$

$$= \boxed{\frac{kQ}{L}\ln\!\left(\frac{\sqrt{x^2+\tfrac14 L^2}+\tfrac12 L}{\sqrt{x^2+\tfrac14 L^2}-\tfrac14 L^2}\right)}$$

(*b*) Factor *x* from the numerator and denominator within the parentheses to obtain:

$$V(x,0) = \frac{kQ}{L}\ln\!\left(\frac{\sqrt{1+\dfrac{L^2}{4x^2}}+\dfrac{L}{2x}}{\sqrt{1+\dfrac{L^2}{4x^2}}-\dfrac{L}{2x}}\right)$$

Use $\ln\!\left(\dfrac{a}{b}\right) = \ln a - \ln b$ to obtain:

$$V(x,0) = \frac{kQ}{L}\left\{\ln\!\left(\sqrt{1+\frac{L^2}{4x^2}}+\frac{L}{2x}\right) - \ln\!\left(\sqrt{1+\frac{L^2}{4x^2}}-\frac{L}{2x}\right)\right\}$$

Let $\varepsilon = \dfrac{L^2}{4x^2}$ and use $(1+\varepsilon)^{1/2} = 1+\tfrac12\varepsilon - \tfrac18\varepsilon^2 + \dots$ to expand $\sqrt{1+\dfrac{L^2}{4x^2}}$:

$$\left(1+\frac{L^2}{4x^2}\right)^{1/2} = 1+\frac{1}{2}\frac{L^2}{4x^2}-\frac{1}{8}\left(\frac{L^2}{4x^2}\right)^2 + \dots \approx 1 \text{ for } |x| \gg L.$$

Substitute for $\left(1+\dfrac{L^2}{4x^2}\right)^{1/2}$ to obtain:

$$V(x,0) = \frac{kQ}{L}\left\{\ln\!\left(1+\frac{L}{2x}\right) - \ln\!\left(1-\frac{L}{2x}\right)\right\}$$

Let $\delta = \dfrac{L}{2x}$ and use $\ln(1+\delta) = \delta - \tfrac{1}{2}\delta^2 + \ldots$ to expand $\ln\left(1 \pm \dfrac{L}{2x}\right)$:

$$\ln\left(1 + \frac{L}{2x}\right) \approx \frac{L}{2x} - \frac{L^2}{4x^2} \text{ and } \ln\left(1 - \frac{L}{2x}\right) \approx -\frac{L}{2x} - \frac{L^2}{4x^2} \text{ for } x \gg L.$$

Substitute for $\ln\left(1 + \dfrac{L}{2x}\right)$ and $\ln\left(1 - \dfrac{L}{2x}\right)$ and simplify to obtain:

$$V(x,0) = \frac{kQ}{L}\left\{\frac{L}{2x} - \frac{L^2}{4x^2} - \left(-\frac{L}{2x} - \frac{L^2}{4x^2}\right)\right\} = \boxed{\frac{kQ}{x}}$$

Because, for $|x| \gg L$, the charge carried by the rod is far enough away from the point of interest to look like a point charge, this result is what we would expect.

53 ••   A disk of radius R has a surface charge distribution given by $\sigma = \sigma_0 r^2/R^2$ where σ_0 is a constant and r is the distance from the center of the disk. (a) Find the total charge on the disk. (b) Find an expression for the electric potential at a distance z from the center of the disk on the axis that passes through the disk's center and is perpendicular to its plane.

Picture the Problem We can find Q by integrating the charge on a ring of radius r and thickness dr from $r = 0$ to $r = R$ and the potential on the axis of the disk by integrating the expression for the potential on the axis of a ring of charge between the same limits.

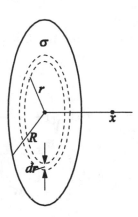

(a) Express the charge dq on a ring of radius r and thickness dr:

$$dq = 2\pi r \sigma dr = 2\pi r\left(\sigma_0 \frac{r^2}{R^2}\right)dr$$

$$= \frac{2\pi\sigma_0}{R^2}r^3 dr$$

Integrate from $r = 0$ to $r = R$ to obtain:

$$Q = \frac{2\pi\sigma_0}{R^2}\int_0^R r^3 dr = \boxed{\tfrac{1}{2}\pi\sigma_0 R^2}$$

(b) Express the potential on the axis of the disk due to a circular element of charge $dq = \dfrac{2\pi\sigma_0}{R^2} r^3 dr$:

$$dV = \frac{kdq}{r'} = \frac{2\pi k\sigma_0}{R^2}\frac{r^3}{\sqrt{x^2+r^2}}dr$$

Integrate from $r = 0$ to $r = R$ to obtain:

$$V = \frac{2\pi k\sigma_0}{R^2}\int_0^R \frac{r^3 dr}{\sqrt{x^2+r^2}} = \boxed{\frac{2\pi k\sigma_0}{R^2}\left(\frac{R^2-2x^2}{3}\sqrt{x^2+R^2}+\frac{2x^3}{3}\right)}$$

59 ••   A circle of radius a is removed from the center of a uniformly charged thin circular disk of radius b. Show that the potential at a point on the central axis of the disk a distance z from its geometrical center is given by
$V(z) = 2\pi k\sigma\left(\sqrt{z^2+b^2}-\sqrt{z^2+a^2}\right)$, where σ is the charge density of the disk.

Picture the Problem We can find the electrostatic potential of the conducting washer by treating it as two disks with equal but opposite charge densities.

The electric potential due to a charged disk of radius R is given by:

$$V(x) = 2\pi k\sigma|x|\left(\sqrt{1+\frac{R^2}{x^2}}-1\right)$$

Superimpose the electrostatic potentials of the two disks with opposite charge densities and simplify to obtain:

$$V(x) = 2\pi k\sigma|x|\left(\sqrt{1+\frac{b^2}{x^2}}-1\right) - 2\pi k\sigma|x|\left(\sqrt{1+\frac{a^2}{x^2}}-1\right)$$

$$= 2\pi k\sigma|x|\left(\left(\sqrt{1+\frac{b^2}{x^2}}-1\right)-\left(\sqrt{1+\frac{a^2}{x^2}}-1\right)\right)$$

$$= 2\pi k\sigma|x|\left(\sqrt{1+\frac{b^2}{x^2}}-\sqrt{1+\frac{a^2}{x^2}}\right) = 2\pi k\sigma|x|\left(\sqrt{\frac{x^2+b^2}{x^2}}-\sqrt{\frac{x^2+a^2}{x^2}}\right)$$

$$= 2\pi k\sigma|x|\left(\frac{\sqrt{x^2+b^2}}{\sqrt{x^2}}-\frac{\sqrt{x^2+a^2}}{\sqrt{x^2}}\right) = 2\pi k\sigma|x|\left(\frac{\sqrt{x^2+b^2}}{|x|}-\frac{\sqrt{x^2+a^2}}{|x|}\right)$$

$$= 2\pi k\sigma\left(\sqrt{x^2+b^2}-\sqrt{x^2+a^2}\right)$$

The charge density σ is given by:

$$\sigma = \frac{Q}{\pi(b^2-a^2)}$$

Substituting for σ yields:

$$V(x) = \boxed{\frac{2kQ}{\left(b^2 - a^2\right)}\left(\sqrt{x^2 + b^2} - \sqrt{x^2 + a^2}\right)}$$

Equipotential Surfaces

61 •• Consider two parallel uniformly charged infinite planes that are equal but oppositely charged. (*a*) What is (are) the shape(s) of the equipotentials in the region between them? Explain your answer. (*b*) What is (are) the shape(s) of the equipotentials in the regions not between them? Explain your answer.

Picture the Problem The two parallel planes, with their opposite charges, are shown in the pictorial representation.

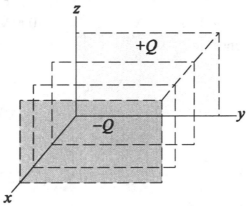

(*a*) Because the electric field between the charged plates is uniform and perpendicular to the plates, the equipotential surfaces are planes parallel to the charged planes.

(*b*) The regions to either side of the two charged planes are equipotential regions, so any surface in either of these regions is an equipotential surface.

63 •• Suppose the cylinder in the Geiger tube in Problem 62 has an inside diameter of 4.00 cm and the wire has a diameter of 0.500 mm. The cylinder is grounded so its potential is equal to zero. (*a*) What is the radius of the equipotential surface that has a potential equal to 500 V? Is this surface closer to the wire or to the cylinder? (*b*) How far apart are the equipotential surfaces that have potentials of 200 and 225 V? (*c*) Compare your result in Part (*b*) to the distance between the two surfaces that have potentials of 700 and 725 V respectively. What does this comparison tell you about the electric field strength as a function of the distance from the central wire?

Picture the Problem If we let the electric potential of the cylinder be zero, then the surface of the central wire is at +1000 V and we can use Equation 23-23 to find the electric potential at any point between the outer cylinder and the central wire.

(*a*) From Equation 23-23 we have:

$$V(r) = 2k\lambda \ln\frac{R_{ref}}{r} \qquad (1)$$

where R_{ref} is the radius of the outer cylinder and r is the distance from the center of the central wire and $r < R_{ref}$.

Solving for $2k\lambda$ yields:

$$2k\lambda = \frac{V(r)}{\ln\dfrac{R_{ref}}{r}}$$

At the surface of the wire, $V = 1000$ V and $r = 0.250$ mm. Hence:

$$2k\lambda = \frac{1000\ \text{V}}{\ln\dfrac{2.00\ \text{cm}}{0.250\ \text{mm}}} = 228.2\ \text{V}$$

and

$$V(r) = (228.2\ \text{V})\ln\frac{2.00\ \text{cm}}{r}$$

Setting $V = 500$ V yields:

$$500\ \text{V} = (228.2\ \text{V})\ln\frac{2.00\ \text{cm}}{r}$$

or

$$\ln\frac{2.00\ \text{cm}}{r} = 2.191 \Rightarrow \frac{2.00\ \text{cm}}{r} = e^{2.191}$$

Solve for r to obtain:

$$r = \frac{2.00\ \text{cm}}{e^{2.191}} = \boxed{0.224\ \text{cm}}, \text{ closer to}$$

the wire.

(*b*) The separation of the equipotential surfaces that have potential values of 200 and 225 V is:

$$\Delta r = \left| r_{225\,\text{V}} - r_{200\,\text{V}} \right| \qquad (2)$$

Solving equation (1) for r yields:

$$r = R_{ref}\, e^{-\frac{V}{2k\lambda}} = (2.00\ \text{cm})e^{-\frac{V}{228.2\ \text{V}}}$$

Substitute for the radii in equation (2), simplify, and evaluate Δr to obtain:

$$\Delta r = \left|(2.00 \text{ cm})e^{-\frac{225 \text{ V}}{228.2 \text{ V}}} - (2.00 \text{ cm})e^{-\frac{200 \text{ V}}{228.2 \text{ V}}}\right| = (2.00 \text{ cm})\left|e^{-\frac{225 \text{ V}}{228.2 \text{ V}}} - e^{-\frac{200 \text{ V}}{228.2 \text{ V}}}\right|$$

$$= \boxed{0.864 \text{ mm}}$$

(c) The distance between the 700 V and the 725 V equipotentials is:

$$\Delta r = (2.00 \text{ cm})\left|e^{-\frac{725 \text{ V}}{228.2 \text{ V}}} - e^{-\frac{700 \text{ V}}{228.2 \text{ V}}}\right|$$

$$= \boxed{0.0966 \text{ mm}}$$

This closer spacing of these two equipotential surfaces was to be expected. Close to the central wire, two equipotential surfaces with the same difference in potential should be closer together to reflect the fact that the higher electric field strength is greater closer to the wire.

Electrostatic Potential Energy

67 •• (a) How much charge is on the surface of an isolated spherical conductor that has a 10.0-cm radius and is charged to 2.00 kV? (b)What is the electrostatic potential energy of this conductor? (Assume the potential is zero far from the sphere.)

Picture the Problem The potential of an isolated spherical conductor is given by $V = kQ/r$, where Q is its charge and r its radius, and its electrostatic potential energy by $U = \frac{1}{2}QV$. We can combine these relationships to find the sphere's electrostatic potential energy.

(a) The potential of the isolated spherical conductor at its surface is related to its radius:

$$V = \frac{kQ}{R} \Rightarrow Q = \frac{RV}{k}$$

where R is the radius of the spherical conductor.

Substitute numerical values and evaluate Q:

$$Q = \frac{(10.0 \text{ cm})(2.00 \text{ kV})}{8.988 \times 10^9 \frac{\text{N} \cdot \text{m}^2}{\text{C}^2}}$$

$$= 22.25 \text{ nC} = \boxed{22.3 \text{ nC}}$$

(b) Express the electrostatic potential energy of the isolated spherical conductor as a function of its charge Q and potential V:

$$U = \tfrac{1}{2}QV$$

Substitute numerical values and evaluate U:

$$U = \tfrac{1}{2}(22.25\ \text{nC})(2.00\ \text{kV}) = \boxed{22.3\ \mu\text{J}}$$

69 •• Four point charges are fixed at the corners of a square centered at the origin. The length of each side of the square is $2a$. The charges are located as follows: $+q$ is at $(-a, +a)$, $+2q$ is at $(+a, +a)$, $-3q$ is at $(+a, -a)$, and $+6q$ is at $(-a, -a)$. A fifth particle that has a mass m and a charge $+q$ is placed at the origin and released from rest. Find its speed when it is a very far from the origin.

Picture the Problem The diagram shows the four point charges fixed at the corners of the square and the fifth charged particle that is released from rest at the origin. We can use conservation of energy to relate the initial potential energy of the particle to its kinetic energy when it is at a great distance from the origin and the electrostatic potential at the origin to express U_i.

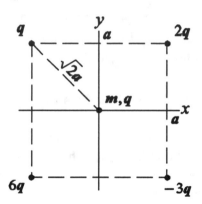

Use conservation of energy to relate the initial potential energy of the particle to its kinetic energy when it is at a great distance from the origin:

$$\Delta K + \Delta U = 0$$
or, because $K_i = U_f = 0$,
$$K_f - U_i = 0$$

Express the initial potential energy of the particle to its charge and the electrostatic potential at the origin:

$$U_i = qV(0)$$

Substitute for K_f and U_i to obtain:

$$\tfrac{1}{2}mv^2 - qV(0) = 0 \Rightarrow v = \sqrt{\frac{2qV(0)}{m}}$$

Express the electrostatic potential at the origin:

$$V(0) = \frac{kq}{\sqrt{2}a} + \frac{2kq}{\sqrt{2}a} + \frac{-3kq}{\sqrt{2}a} + \frac{6kq}{\sqrt{2}a}$$

$$= \frac{6kq}{\sqrt{2}a}$$

Substitute for $V(0)$ and simplify to obtain:

$$v = \sqrt{\frac{2q}{m}\left(\frac{6kq}{\sqrt{2}a}\right)} = \boxed{q\sqrt{\frac{6\sqrt{2}k}{ma}}}$$

General Problems

73 •   Two positive point charges, each have a charge of $+q$, and are fixed on the y-axis at $y = +a$ and $y = -a$. (a) Find the electric potential at any point on the x-axis. (b) Use your result in Part (a) to find the electric field at any point on the x-axis.

Picture the Problem The potential V at any point on the x axis is the sum of the Coulomb potentials due to the two point charges. Once we have found V, we can use $\vec{E} = -(\partial V_x / \partial x)\hat{i}$ to find the electric field at any point on the x axis.

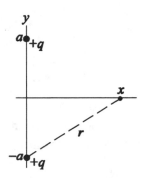

(a) Express the potential due to a system of point charges:

$$V = \sum_i \frac{kq_i}{r_i}$$

Substitute to obtain:

$$V(x) = V_{\text{charge at } +a} + V_{\text{charge at } -a}$$

$$= \frac{kq}{\sqrt{x^2 + a^2}} + \frac{kq}{\sqrt{x^2 + a^2}}$$

$$= \boxed{\frac{2kq}{\sqrt{x^2 + a^2}}}$$

(b) The electric field at any point on the x axis is given by:

$$\vec{E}(x) = -\frac{\partial V_x}{\partial x}\hat{i} = -\frac{d}{dx}\left[\frac{2kq}{\sqrt{x^2 + a^2}}\right]\hat{i}$$

$$= \boxed{\frac{2kqx}{\left(x^2 + a^2\right)^{3/2}}\hat{i}}$$

75 ••   Two infinitely long parallel wires have a uniform charge per
unit length λ and $-\lambda$ respectively. The wires are parallel with the z-axis. The
positively charged wire intersects the x-axis at $x = -a$, and the negatively charged
wire intersects the x-axis at $x = +a$. (a) Choose the origin as the reference point
where the potential is zero, and express the potential at an arbitrary point (x, y) in
the xy plane in terms of x, y, λ, and a. Use this expression to solve for the
potential everywhere on the y axis. (b) Using $a = 5.00$ cm and $\lambda = 5.00$ nC/m,
obtain the equation for the equipotential in the xy plane that passes through the
point $x = \frac{1}{4}a$, $y = 0$. (c) Use a **spreadsheet** program to plot the equipotential found
in part (b).

Picture the Problem The geometry of the wires is shown below. The potential at
the point whose coordinates are (x, y) is the sum of the potentials due to the
charge distributions on the wires.

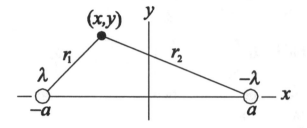

(a) Express the potential at the
point whose coordinates are (x, y):

$$V(x, y) = V_{\text{wire at } -a} + V_{\text{wire at } a}$$

$$= 2k\lambda \ln\left(\frac{r_{\text{ref}}}{r_1}\right) + 2k(-\lambda)\ln\left(\frac{r_{\text{ref}}}{r_2}\right)$$

$$= 2k\lambda\left[\ln\left(\frac{r_{\text{ref}}}{r_1}\right) - \ln\left(\frac{r_{\text{ref}}}{r_2}\right)\right]$$

$$= \frac{\lambda}{2\pi\,\epsilon_0}\ln\left(\frac{r_2}{r_1}\right)$$

where $V(0) = 0$.

Because $r_1 = \sqrt{(x+a)^2 + y^2}$ and
$r_2 = \sqrt{(x-a)^2 + y^2}$:

$$\boxed{V(x, y) = \frac{\lambda}{2\pi\,\epsilon_0}\ln\left(\frac{\sqrt{(x-a)^2 + y^2}}{\sqrt{(x+a)^2 + y^2}}\right)}$$

On the y-axis, $x = 0$ and:

$$V(0, y) = \frac{\lambda}{2\pi\,\epsilon_0}\ln\left(\frac{\sqrt{a^2 + y^2}}{\sqrt{a^2 + y^2}}\right)$$

$$= \frac{\lambda}{2\pi\,\epsilon_0}\ln(1) = \boxed{0}$$

(b) Evaluate the potential at $\left(\frac{1}{4}a,0\right)=(1.25\,\text{cm},0)$:

$$V\left(\tfrac{1}{4}a,0\right)=\frac{\lambda}{2\pi\,\epsilon_0}\ln\left(\frac{\sqrt{\left(\frac{1}{4}a-a\right)^2}}{\sqrt{\left(\frac{1}{4}a+a\right)^2}}\right)$$

$$=\frac{\lambda}{2\pi\,\epsilon_0}\ln\left(\frac{3}{5}\right)$$

Equate $V(x,y)$ and $V\left(\frac{1}{4}a,0\right)$:

$$\frac{3}{5}=\frac{\sqrt{(x-5)^2+y^2}}{\sqrt{(x+5)^2+y^2}}$$

Solve for y to obtain:

$$y=\boxed{\pm\sqrt{21.25x-x^2-25}}$$

(c) A spreadsheet program to plot $y=\pm\sqrt{21.25x-x^2-25}$ is shown below. The formulas used to calculate the quantities in the columns are as follows:

Cell	Content/Formula	Algebraic Form
A2	1.25	$\frac{1}{4}a$
A3	A2 + 0.05	$x+\Delta x$
B2	SQRT(21.25*A2 − A2^2 − 25)	$y=\sqrt{21.25x-x^2-25}$
B4	−B2	$y=-\sqrt{21.25x-x^2-25}$

	A	B	C
	x	y_{pos}	y_{neg}
2	1.25	0.00	0.00
3	1.30	0.97	−0.97
4	1.35	1.37	−1.37
5	1.40	1.67	−1.67
6	1.45	1.93	−1.93
7	1.50	2.15	−2.15
370	19.65	2.54	−2.54
371	19.70	2.35	−2.35
372	19.75	2.15	−2.15
373	19.80	1.93	−1.93
374	19.85	1.67	−1.67
375	19.90	1.37	−1.37
376	19.95	0.97	−0.97

The following graph shows the equipotential curve in the xy plane for

$$V\left(\tfrac{1}{4}a,0\right) = \frac{\lambda}{2\pi \,\epsilon_0}\ln\!\left(\frac{3}{5}\right).$$

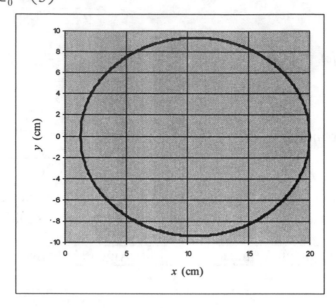

87 •••   Configuration A consists of two point particles, one particle has a charge of $+q$ and is on the x axis at $x = +d$ and the other particle has a charge of $-q$ and is at $x = -d$ (Figure 23-36a). (a) Assuming the potential is zero at large distances from these charged particles, show that the potential is also zero everywhere on the $x = 0$ plane. (b) Configuration B consists of a flat metal plate of infinite extent and a point particle located a distance d from the plate (Figure 23-36b) The point particle has a charge equal to $+q$ and the plate is grounded. (Grounding the plate forces its potential to equal zero.) Choose the line perpendicular to the plate and through the point charge as the x axis, and choose the origin at the surface of the plate nearest the particle. (These choices put the particle on the x axis at $x = +d$.) For configuration B, the electric potential is zero both at all points in the half-space $x \geq 0$ that are very far from the particle and at all points on the $x = 0$ plane—just as was the case for configuration A. (c) A theorem, called the *uniqueness theorem*, implies that throughout the half-space $x \geq 0$ the potential function V—and thus the electric field $\vec{E}$—for the two configurations are identical. Using this result, obtain the electric field $\vec{E}$ at every point in the $x = 0$ plane in the configuration B. (The uniqueness theorem tells us that in configuration B the electric field at each point in the $x = 0$ plane is the same as it is in configuration A.) Use this result to find the surface charge density σ at each point in the conducting plane (in configuration B).

Picture the Problem We can use the relationship between the potential and the electric field to show that this arrangement is equivalent to replacing the plane by a point charge of magnitude $-q$ located a distance d beneath the plane. In (b) we can first find the field at the plane surface and then use $\sigma = \epsilon_0 E$ to find the surface charge density. In (c) the work needed to move the charge to a point $2d$ away from the plane is the product of the potential difference between the points at distances $2d$ and $3d$ from $-q$ multiplied by the separation Δx of these points.

(a) The potential anywhere on the plane is 0 and the electric field is perpendicular to the plane in both configurations, so they must give the same potential everywhere in the xy plane. Also, because the net charge is zero, the potential at infinity is zero.

(b) The surface charge density is given by:

$$\sigma = \epsilon_0 E \qquad (1)$$

At any point on the plane, the electric field points in the negative x direction and has magnitude:

$$E = \frac{kq}{d^2 + r^2} \cos\theta$$

where θ is the angle between the horizontal and a vector pointing from the positive charge to the point of interest on the xz plane and r is the distance along the plane from the origin (that is, directly to the left of the charge).

Because $\cos\theta = \dfrac{d}{\sqrt{d^2 + r^2}}$:

$$E = \frac{kq}{d^2 + r^2} \frac{d}{\sqrt{d^2 + r^2}} = \frac{kqd}{\left(d^2 + r^2\right)^{3/2}}$$

$$= \frac{qd}{4\pi \epsilon_0 \left(d^2 + r^2\right)^{3/2}}$$

Substitute for E in equation (1) and simplify to obtain:

$$\sigma = \boxed{\frac{qd}{4\pi\left(d^2 + r^2\right)^{3/2}}}$$

91 ••• Show that the total work needed to assemble a uniformly charged sphere that has a total charge of Q and radius R is given by $3Q^2/(20\pi \epsilon_0 R)$. Energy conservation tells us that this result is the same as the resulting electrostatic potential energy of the sphere. *Hint: Let ρ be the charge density of the sphere that has charge Q and radius R. Calculate the work dW to bring in charge dq from infinity to the surface of a uniformly charged sphere that has radius r (r < R) and charge density ρ. (No additional work is required to smear dq throughout a spherical shell of radius r, thickness dr, and charge density ρ. Why?)*

Picture the Problem We can use the hint to derive an expression for the electrostatic potential energy dU required to bring in a layer of charge of thickness dr and then integrate this expression from $r = 0$ to R to obtain an expression for the required work.

If we build up the sphere in layers, then at a given radius r the net charge on the sphere will be given by:

$$Q(r) = Q\left(\frac{r}{R}\right)^3$$

When the radius of the sphere is r, the potential relative to infinity is:

$$V(r) = \frac{Q(r)}{4\pi \epsilon_0 \, r} = \frac{Q}{4\pi \epsilon_0} \frac{r^2}{R^3}$$

Express the work dW required to bring in charge dQ from infinity to the surface of a uniformly charged sphere of radius r:

$$dW = dU = V(r)dQ$$

$$= \frac{Q}{4\pi \epsilon_0} \frac{r^2}{R^3}\left(4\pi r^2 \frac{3Q}{4\pi R^3} dr\right)$$

$$= \frac{3Q^2}{4\pi \epsilon_0 R^6} r^4 dr$$

Integrate dW from 0 to R to obtain:

$$W = U = \frac{3Q^2}{4\pi \epsilon_0 R^6}\int_0^R r^4 dr$$

$$= \frac{3Q^2}{4\pi \epsilon_0 R^6}\left[\frac{r^5}{5}\right]_0^R = \frac{3Q^2}{20\pi \epsilon_0 R}$$

$$= \boxed{\frac{3Q^2}{20\pi \epsilon_0 R}}$$

93 ••• (*a*) Consider a uniformly charged sphere that has radius R and charge Q and is composed of an incompressible fluid, such as water. If the sphere fissions (splits) into two halves of equal volume and equal charge, and if these halves stabilize into uniformly charged spheres, what is the radius R' of each? (*b*) Using the expression for potential energy shown in Problem 90, calculate the change in the total electrostatic potential energy of the charged fluid. Assume that the spheres are separated by a large distance.

Picture the Problem Because the post-fission volumes of the fission products are equal, we can express the post-fission radii in terms of the radius of the pre-fission sphere.

(*a*) Relate the initial volume V of the uniformly charged sphere to the volumes V' of the fission products:

$$V = 2V'$$

Substitute for V and V':

$$\tfrac{4}{3}\pi R^3 = 2\left(\tfrac{4}{3}\pi R'^3\right)$$

Solving for R' yields:

$$R' = \frac{1}{\sqrt[3]{2}} R = \boxed{0.794R}$$

(b) Express the difference ΔE in the total electrostatic energy as a result of fissioning:

$$\Delta E = E' - E$$

From Problem 91 we have:

$$E = \frac{3Q^2}{20\pi \, \epsilon_0 \, R}$$

After fissioning:

$$E' = 2\left(\frac{3Q'^2}{20\pi \, \epsilon_0 \, R'} \right) = 2\left[\frac{3\left(\frac{1}{2}Q\right)^2}{20\pi \, \epsilon_0 \, \frac{1}{\sqrt[3]{2}} R} \right]$$

$$= \frac{\sqrt[3]{2}}{2}\left(\frac{3Q^2}{20\pi \, \epsilon_0 \, R} \right) = 0.630E$$

Substitute for E and E' to obtain:

$$\Delta E = 0.630E - E = \boxed{-0.370E}$$

Chapter 24
Capacitance

Conceptual Problems

5 •   A parallel-plate capacitor is connected to a battery. The space between the two plates is empty. If the separation between the capacitor plates is tripled while the capacitor remains connected to the battery, what is the ratio of the final stored energy to the initial stored energy?

Determine the Concept The energy stored in a capacitor is given by $U = \frac{1}{2}QV$ and the capacitance of a parallel-plate capacitor by $C = \epsilon_0 A/d$. We can combine these relationships, using the definition of capacitance and the condition that the potential difference across the capacitor is constant, to express U as a function of d.

Express the energy stored in the capacitor:

$$U = \tfrac{1}{2}QV \qquad\qquad (1)$$

Use the definition of capacitance to express the charge of the capacitor:

$$Q = CV$$

Express the capacitance of a parallel-plate capacitor in terms of the separation d of its plates:

$$C = \frac{\epsilon_0 A}{d}$$

where A is the area of one plate.

Substituting for Q and C in equation (1) yields:

$$U = \frac{\epsilon_0 A V^2}{2d}$$

Because $U \propto \dfrac{1}{d}$, tripling the separation of the plates will reduce the energy stored in the capacitor to one-third its previous value. Hence the ratio of the final stored energy to the initial stored energy is $\boxed{1/3}$.

9 •   A dielectric is inserted between the plates of a parallel-plate capacitor, completely filling the region between the plates. Air initially filled the region between the two plates. The capacitor was connected to a battery during the entire process. True or false:

(a)   The capacitance value of the capacitor increases as the dielectric is inserted between the plates.

(b)   The charge on the capacitor plates decreases as the dielectric is inserted between the plates.
(c)   The electric field between the plates does not change as the dielectric is inserted between the plates.
(d)   The energy storage of the capacitor decreases as the dielectric is inserted between the plates.

Determine the Concept The capacitance of the capacitor is given by $C = \dfrac{\kappa\,\epsilon_0\,A}{d}$, the charge on the capacitor is given by $Q = CV$, and the energy stored in the capacitor is given by $U = \tfrac{1}{2}CV^2$.

(a) True. As the dielectric material is inserted, κ increases from 1 (air) to its value for the given dielectric material.

(b) False. Because $Q = CV$, and C increases, Q must increase.

(c) True. $E = V/d$, where d is the plate separation.

(d) False. The energy storage of a capacitor is independent of the presence of dielectric and is given by $U = \tfrac{1}{2}QV$.

11 ••   (a) Two identical capacitors are connected in parallel. This combination is then connected across the terminals of a battery. How does the total energy stored in the parallel combination of the two capacitors compare to the total energy stored if just one of the capacitors were connected across the terminals of the same battery? (b) Two identical capacitors that have been discharged are connected in series. This combination is then connected across the terminals of a battery. How does the total energy stored in the series combination of the two capacitors compare to the total energy stored if just one of the capacitors were connected across the terminals of the same battery?

Picture the Problem The energy stored in a capacitor whose capacitance is C and across which there is a potential difference V is given by $U = \tfrac{1}{2}CV^2$. Let C_0 represent the capacitance of the each of the two identical capacitors.

(a) The energy stored in the parallel system is given by:

$$U_{\text{parallel}} = \tfrac{1}{2}C_{\text{eq}}V^2$$

When the capacitors are connected in parallel, their equivalent capacitance is:

$$C_{\text{parallel}} = C_0 + C_0 = 2C_0$$

Substituting for C_{eq} and simplifying yields:

$$U_{parallel} = \tfrac{1}{2}(2C_0)V^2 = C_0V^2 \qquad (1)$$

If just one capacitor is connected to the same battery the stored energy is:

$$U_{1\,capacitor} = \tfrac{1}{2}C_0V^2 \qquad\qquad (2)$$

Dividing equation (1) by equation (2) and simplifying yields:

$$\frac{U_{parallel}}{U_{1\,capacitor}} = \frac{C_0V^2}{\tfrac{1}{2}C_0V^2} = 2$$

or

$$U_{parallel} = \boxed{2U_{1\,capacitor}}$$

(b) The energy stored in the series system is given by:

$$U_{series} = \tfrac{1}{2}C_{eq}V^2$$

When the capacitors are connected in series, their equivalent capacitance is:

$$C_{series} = \tfrac{1}{2}C_0$$

Substituting for C_{eq} and simplifying yields:

$$U_{series} = \tfrac{1}{2}\left(\tfrac{1}{2}C_0\right)V^2 = \tfrac{1}{4}C_0V^2 \quad (3)$$

Dividing equation (3) by equation (2) and simplifying yields:

$$\frac{U_{series}}{U_{1\,capacitor}} = \frac{\tfrac{1}{4}C_0V^2}{\tfrac{1}{2}C_0V^2} = \tfrac{1}{2}$$

or

$$U_{series} = \boxed{\tfrac{1}{2}U_{1\,capacitor}}$$

Estimation and Approximation

13 •• Disconnect the coaxial cable from a television or other device and estimate the diameter of the inner conductor and the diameter of the shield. Assume a plausible value (see Table 24–1) for the dielectric constant of the dielectric separating the two conductors and estimate the capacitance per unit length of the cable.

Picture the Problem The outer diameter of a "typical" coaxial cable is about 5 mm, while the inner diameter is about 1 mm. From Table 24-1 we see that a reasonable range of values for κ is 3-5. We can use the expression for the capacitance of a cylindrical capacitor to estimate the capacitance per unit length of a coaxial cable.

The capacitance of a cylindrical dielectric-filled capacitor is given by:

$$C = \frac{2\pi\kappa\,\epsilon_0\,L}{\ln\left(\dfrac{R_2}{R_1}\right)}$$

where L is the length of the capacitor, R_1 is the radius of the inner conductor, and R_2 is the radius of the second (outer) conductor.

Divide both sides by L to obtain an expression for the capacitance per unit length of the cable:

$$\frac{C}{L} = \frac{2\pi\kappa\,\epsilon_0}{\ln\left(\dfrac{R_2}{R_1}\right)} = \frac{\kappa}{2k\ln\left(\dfrac{R_2}{R_1}\right)}$$

If $\kappa = 3$:

$$\frac{C}{L} = \frac{3}{2\left(8.988\times10^9\,\text{N}\cdot\text{m}^2/\text{C}^2\right)\ln\left(\dfrac{2.5\,\text{mm}}{0.5\,\text{mm}}\right)} \approx 0.1\,\text{nF/m}$$

If $\kappa = 5$:

$$\frac{C}{L} = \frac{5}{2\left(8.988\times10^9\,\text{N}\cdot\text{m}^2/\text{C}^2\right)\ln\left(\dfrac{2.5\,\text{mm}}{0.5\,\text{mm}}\right)} \approx 0.2\,\text{nF/m}$$

A reasonable range of values for C/L, corresponding to $3 \le \kappa \le 5$, is:

$$\boxed{0.1\,\text{nF/m} \le \frac{C}{L} \le 0.2\,\text{nF/m}}$$

15 ••   Estimate the capacitance of the Leyden jar shown in the Figure 24-34. The figure of a man is one-tenth the height of an average man.

Picture the Problem Modeling the Leyden jar as a parallel-plate capacitor, we can use the equation for the capacitance of a dielectric-filled parallel-plate capacitor that relates its capacitance to the area A of its plates and their separation (the thickness of the glass) d to estimate the capacitance of the jar. See Table 24-1 for the dielectric constants of various materials.

The capacitance of a dielectric-filled parallel-plate capacitor is given by:

$$C = \frac{\kappa\,\epsilon_0\,A}{d}$$

where κ is the dielectric constant.

Let the plate area be the sum of the area of the lateral surface of the jar and its base:

$$A = A_{lateral} + A_{base} = 2\pi Rh + \pi R^2$$
$$\text{area}$$

where h is the height of the jar and R is its inside radius.

Substitute for A and simplify to obtain:

$$C = \frac{\kappa \epsilon_0 \left(2\pi Rh + \pi R^2\right)}{d}$$
$$= \frac{\pi \kappa \epsilon_0 \, R(2h + R)}{d}$$

If the glass of the Leyden jar is Bakelite of thickness 2.0 mm and the radius and height of the jar are 4.0 cm and 40 cm, respectively, then:

$$C = \frac{\pi(4.9)(4.0\,\text{cm})\left(8.854\times10^{-12}\,\dfrac{C^2}{N\cdot m^2}\right)\left[2(40\,\text{cm}) + 4.0\,\text{cm}\right]}{2.0\,\text{mm}} = \boxed{2.3\,\text{nF}}$$

The Storage of Electrical Energy

19 • (a) The potential difference between the plates of a 3.00-μF capacitor is 100 V. How much energy is stored in the capacitor? (b) How much additional energy is required to increase the potential difference between the plates from 100 V to 200 V?

Picture the Problem Of the three equivalent expressions for the energy stored in a charged capacitor, the one that relates U to C and V is $U = \frac{1}{2}CV^2$.

(a) Express the energy stored in the capacitor as a function of C and V:

$$U = \tfrac{1}{2}CV^2$$

Substitute numerical values and evaluate U:

$$U = \tfrac{1}{2}(3.00\,\mu F)(100\,V)^2 = \boxed{15.0\,\text{mJ}}$$

(b) Express the additional energy required as the difference between the energy stored in the capacitor at 200 V and the energy stored at 100 V:

$$\Delta U = U(200\,V) - U(100\,V)$$
$$= \tfrac{1}{2}(3.00\,\mu F)(200\,V)^2 - 15.0\,\text{mJ}$$
$$= \boxed{45.0\,\text{mJ}}$$

23 ••   An air-gap parallel-plate capacitor that has a plate area of $2.00 \, \text{m}^2$ and a separation of 1.00 mm is charged to 100 V. (*a*) What is the electric field between the plates? (*b*) What is the electric energy density between the plates? (*c*) Find the total energy by multiplying your answer from Part (*b*) by the volume between the plates. (*d*) Determine the capacitance of this arrangement. (*e*) Calculate the total energy from $U = \frac{1}{2}CV^2$, and compare your answer with your result from Part (*c*).

Picture the Problem Knowing the potential difference between the plates, we can use $E = V/d$ to find the electric field between them. The energy per unit volume is given by $u = \frac{1}{2}\epsilon_0 E^2$ and we can find the capacitance of the parallel-plate capacitor using $C = \epsilon_0 A/d$.

(*a*) Express the electric field between the plates in terms of their separation and the potential difference between them:

$$E = \frac{V}{d} = \frac{100 \, \text{V}}{1.00 \, \text{mm}} = \boxed{100 \, \text{kV/m}}$$

(*b*) Express the energy per unit volume in an electric field:

$$u = \frac{1}{2}\epsilon_0 E^2$$

Substitute numerical values and evaluate *u*:

$$u = \frac{1}{2}\left(8.854 \times 10^{-12} \, \frac{\text{C}^2}{\text{N} \cdot \text{m}^2}\right)(100 \, \text{kV/m})^2$$

$$= 44.27 \, \text{mJ/m}^3 = \boxed{44.3 \, \text{mJ/m}^3}$$

(*c*) The total energy is given by:

$$U = uV = uAd$$

$$= (44.27 \, \text{mJ/m}^3)(2.00 \, \text{m}^2)(1.00 \, \text{mm})$$

$$= \boxed{88.5 \, \mu\text{J}}$$

(*d*) The capacitance of a parallel-plate capacitor is given by:

$$C = \frac{\epsilon_0 A}{d}$$

Substitute numerical values and evaluate *C*:

$$C = \frac{\left(8.854 \times 10^{-12} \, \frac{\text{C}^2}{\text{N} \cdot \text{m}^2}\right)(2.00 \, \text{m}^2)}{1.00 \, \text{mm}}$$

$$= 17.71 \, \text{nF} = \boxed{17.7 \, \text{nF}}$$

(*e*) The total energy is given by:

$$U = \frac{1}{2}CV^2$$

Substitute numerical values and evaluate U:

$$U = \tfrac{1}{2}(17.71\,\text{nF})(100\,\text{V})^2$$

$$= \boxed{88.5\,\mu\text{J, in agreement with } (c).}$$

Spherical Capacitors

49 ••   A spherical capacitor consists of a thin spherical shell that has a radius R_1 and a thin, concentric spherical shell that has a radius R_2, where $R_2 > R_1$. (a) Show that the capacitance is given by $C = 4\pi\epsilon_0 R_1 R_2/(R_2 - R_1)$. (b) Show that when the radii of the shells are nearly equal, the capacitance approximately is given by the expression for the capacitance of a parallel-plate capacitor, $C = \epsilon_0 A/d$, where A is the area of the sphere and $d = R_2 - R_1$.

Picture the Problem We can use the definition of capacitance and the expression for the potential difference between charged concentric spherical shells to show that $C = 4\pi\epsilon_0\, R_1 R_2 /(R_2 - R_1)$.

(a) Using its definition, relate the capacitance of the concentric spherical shells to their charge Q and the potential difference V between their surfaces:

$$C = \frac{Q}{V}$$

Express the potential difference between the conductors:

$$V = kQ\left(\frac{1}{R_1} - \frac{1}{R_2}\right) = kQ\frac{R_2 - R_1}{R_1 R_2}$$

Substitute for V and simplify to obtain:

$$C = \frac{Q}{kQ\dfrac{R_2 - R_1}{R_1 R_2}} = \frac{R_1 R_2}{k(R_2 - R_1)}$$

$$= \boxed{\frac{4\pi\epsilon_0\, R_1 R_2}{R_2 - R_1}}$$

(b) Because $R_2 = R_1 + d$:

$$R_1 R_2 = R_1(R_1 + d)$$
$$= R_1^2 + R_1 d$$
$$\approx R_1^2 = R^2$$

because d is small.

Substitute to obtain:

$$C \approx \frac{4\pi\epsilon_0\, R^2}{d} = \boxed{\frac{\epsilon_0 A}{d}}$$

Disconnected and Reconnected Capacitors

53 ••   A 100-pF capacitor and a 400-pF capacitor are both charged to 2.00 kV. They are then disconnected from the voltage source and are connected together, positive plate to negative plate and negative plate to positive plate. (*a*) Find the resulting potential difference across each capacitor. (*b*) Find the energy dissipated when the connections are made.

Picture the Problem (*a*) Just after the two capacitors are disconnected from the voltage source, the 100-pF capacitor carries a charge of 200 nC and the 400-pF capacitor carries a charge of 800 nC. After switches S_1 and S_2 in the circuit are closed, the capacitors are in parallel between points *a* and *b*, and the equivalent capacitance of the system is $C_{eq} = C_{100} + C_{400}$. The plates to the right in the diagram below form a single conductor with a charge of 600 nC, and the plates to the left form a conductor with charge $-Q = -600$ nC. The potential difference across each capacitor is $V = Q/C_{eq}$. In Part (*b*) we can find the energy dissipated when the connections are made by subtracting the energy stored in the system after they are connected from the energy stored in the system before they are connected.

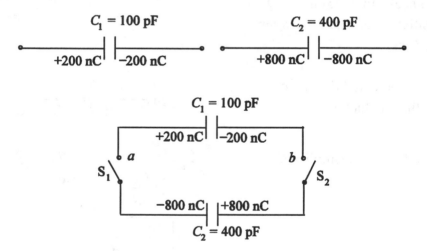

(*a*) When the switches are closed and the capacitors are connected together, their initial charges redistribute and the final charge on the two-capacitor system is 600 nC and the equivalent capacitance is 500 pF:

The potential difference across each capacitor is the potential difference across the equivalent capacitor:

$$V = \frac{Q}{C_{eq}} = \frac{600 \text{ nC}}{500 \text{ pF}} = \boxed{1.20 \text{ kV}}$$

(b) The energy dissipated when the capacitors are connected is the difference between the energy stored after they are connected and the energy stored before they were connected:

$$U_{\text{dissipated}} = U_{\text{before}} - U_{\text{after}} \qquad (1)$$

U_{before} is given by:

$$U_{\text{before}} = U_1 + U_2$$
$$= \tfrac{1}{2}Q_1V_1 + \tfrac{1}{2}Q_2V_2$$
$$= \tfrac{1}{2}(Q_1 + Q_2)V_i$$

where V_i is the charging voltage.

U_{after} is given by:

$$U_{\text{after}} = \tfrac{1}{2}QV$$

where Q is the total charge stored after the capacitors have been connected and V is the voltage found in Part (a).

Substitute for U_{before} and U_{after} in equation (1) and simplify to obtain:

$$U_{\text{dissipated}} = \tfrac{1}{2}(Q_1 + Q_2)V_i - \tfrac{1}{2}QV$$

Substitute numerical values and evaluate $U_{\text{dissipated}}$:

$$U_{\text{dissipated}} = \tfrac{1}{2}(200 \text{ nC} + 800 \text{ nC})(2.00 \text{ kV}) - \tfrac{1}{2}(600 \text{ nC})(1.20 \text{ kV}) = \boxed{640 \ \mu J}$$

59 ••   Capacitors 1, 2 and 3, have capacitances equal to 2.00 μF, 4.00 μF, and 6.00 μF, respectively. The capacitors are connected in parallel, and the parallel combination is connected across the terminals of a 200-V source. The capacitors are then disconnected from both the voltage source and each other, and are connected to three switches as shown in Figure 24-42. (a) What is the potential difference across each capacitor when switches S_1 and S_2 are closed but switch S_3 remains open? (b) After switch S_3 is closed, what is the final charge on the leftmost plate of each capacitor? (c) Give the final potential difference across each capacitor after switch S_3 is closed.

Picture the Problem Let lower case qs refer to the charges before S_3 is closed and upper case Qs refer to the charges after this switch is closed. We can use conservation of charge to relate the charges on the capacitors before S_3 is closed to their charges when this switch is closed. We also know that the sum of the potential differences around the circuit when S_3 is closed must be zero and can use this to obtain a fourth equation relating the charges on the capacitors after the switch is closed to their capacitances. Solving these equations simultaneously will yield the charges Q_1, Q_2, and Q_3. Knowing these charges, we can use the definition of capacitance to find the potential difference across each of the capacitors.

(a) With S_1 and S_2 closed, but S_3 open, the charges on and the potential differences across the capacitors do not change. Hence:

$$V_1 = V_2 = V_3 = \boxed{200\,\text{V}}$$

(b) When S_3 is closed, the charges can redistribute; express the conditions on the charges that must be satisfied as a result of this redistribution:

$$q_2 - q_1 = Q_2 - Q_1,$$
$$q_3 - q_2 = Q_3 - Q_2,$$
and
$$q_1 - q_3 = Q_1 - Q_3.$$

Express the condition on the potential differences that must be satisfied when S_3 is closed:

$$V_1 + V_2 + V_3 = 0$$
where the subscripts refer to the three capacitors.

Use the definition of capacitance to eliminate the potential differences:

$$\frac{Q_1}{C_1} + \frac{Q_2}{C_2} + \frac{Q_3}{C_3} = 0 \qquad (1)$$

Use the definition of capacitance to find the initial charge on each capacitor:

$$q_1 = C_1 V = (2.00\,\mu\text{F})(200\,\text{V}) = 400\,\mu\text{C},$$
$$q_2 = C_2 V = (4.00\,\mu\text{F})(200\,\text{V}) = 800\,\mu\text{C},$$
and
$$q_3 = C_3 V = (6.00\,\mu\text{F})(200\,\text{V}) = 1200\,\mu\text{C}$$

Let $q = q_1$. Then:

$$q_2 = 2Q \text{ and } q_3 = 3Q$$

Express Q_2 and Q_3 in terms of Q_1 and Q:

$$Q_2 = Q + Q_1 \qquad (2)$$
and
$$Q_3 = Q_1 + 2Q \qquad (3)$$

Substitute in equation (1) to obtain:

$$\frac{Q_1}{C_1} + \frac{Q + Q_1}{C_2} + \frac{Q_1 + 2Q}{C_3} = 0$$
or
$$\frac{Q_1}{2.00\,\mu\text{F}} + \frac{Q + Q_1}{4.00\,\mu\text{F}} + \frac{Q_1 + 2Q}{6.00\,\mu\text{F}} = 0$$

Solve for and evaluate Q_1 to obtain:

$$Q_1 = -\tfrac{7}{11}Q = -\tfrac{7}{11}(400\,\mu\text{C}) = \boxed{-255\,\mu\text{C}}$$

Substitute in equation (2) to obtain:

$$Q_2 = 400\,\mu C - 255\,\mu C = \boxed{145\,\mu C}$$

Substitute in equation (3) to obtain:

$$Q_3 = -255\,\mu C + 2(400\,\mu C) = \boxed{545\,\mu C}$$

(c) Use the definition of capacitance to find the potential difference across each capacitor with S_3 closed:

$$V_1 = \frac{Q_1}{C_1} = \frac{-255\,\mu C}{2.00\,\mu F} = \boxed{-127\,V},$$

$$V_2 = \frac{Q_2}{C_2} = \frac{145\,\mu C}{4.00\,\mu F} = \boxed{36.4\,V},$$

and

$$V_3 = \frac{Q_3}{C_3} = \frac{545\,\mu C}{6.00\,\mu F} = \boxed{90.9\,V}$$

General Problems

79 ••   A parallel combination of two identical 2.00-μF parallel-plate capacitors (no dielectric is in the space between the plates) is connected to a 100-V battery. The battery is then removed and the separation between the plates of one of the capacitors is doubled. Find the charge on the positively charged plate of each of the capacitors.

Picture the Problem When the battery is removed, after having initially charged both capacitors, and the separation of one of the capacitors is doubled, the charge is redistributed subject to the condition that the total charge remains constant; that is, $Q = Q_1 + Q_2$ where Q is the initial charge on both capacitors and Q_2 is the charge on the capacitor whose plate separation has been doubled. We can use the conservation of charge during the plate separation process and the fact that, because the capacitors are in parallel, they share a common potential difference.

Find the equivalent capacitance of the two 2.00-μF parallel-plate capacitors connected in parallel:

$$C_{eq} = 2.00\,\mu F + 2.00\,\mu F = 4.00\,\mu F$$

Use the definition of capacitance to find the charge on the equivalent capacitor:

$$Q = C_{eq}V = (4.00\,\mu F)(100\,V) = 400\,\mu C$$

Relate this total charge to charges distributed on capacitors 1 and 2 when the battery is removed and the separation of the plates of capacitor 2 is doubled:

$$Q = Q_1 + Q_2 \qquad\qquad (1)$$

Because the capacitors are in parallel:

$$V_1 = V_2 \text{ and } \frac{Q_1}{C_1} = \frac{Q_2}{C_2'} = \frac{Q_2}{\frac{1}{2}C_2} = \frac{2Q_2}{C_2}$$

Solve for Q_1 to obtain:

$$Q_1 = 2\left(\frac{C_1}{C_2}\right)Q_2 \qquad (2)$$

Substitute equation (2) in equation (1) and solve for Q_2 to obtain:

$$Q_2 = \frac{Q}{2(C_1/C_2)+1}$$

Substitute numerical values and evaluate Q_2:

$$Q_2 = \frac{400\,\mu C}{2(2.00\,\mu F/2.00\,\mu F)+1} = \boxed{133\,\mu C}$$

Substitute numerical values in equation (1) or equation (2) and evaluate Q_1:

$$Q_1 = \boxed{267\,\mu C}$$

85 •••   An electrically isolated capacitor that has charge Q on its positively charged plate is partly filled with a dielectric substance as shown in Figure 24-51. The capacitor consists of two rectangular plates that have edge lengths a and b and are separated by distance d. The dielectric is inserted into the gap a distance x. (*a*) What is the energy stored in the capacitor? *Hint: the capacitor can be modeled as two capacitors connected in parallel.* (*b*) Because the energy of the capacitor decreases as x increases, the electric field must be doing work on the dielectric, meaning that there must be an electric force pulling it in. Calculate this force by examining how the stored energy varies with x. (*c*) Express the force in terms of the capacitance and potential difference V between the plates. (*d*) From where does this force originate?

Picture the Problem We can model this capacitor as the equivalent of two capacitors connected in parallel, one with an air gap and other filled with a dielectric of constant κ. Let the numeral 1 denote the capacitor with the dielectric material whose constant is κ and the numeral 2 the air-filled capacitor.

(*a*) Using the hint, express the energy stored in the capacitor as a function of the equivalent capacitance C_{eq}:

$$U = \frac{1}{2}\frac{Q^2}{C_{eq}}$$

The capacitances of the two capacitors are:

$$C_1 = \frac{\kappa\,\epsilon_0\,ax}{d} \text{ and } C_2 = \frac{\epsilon_0\,a(a-x)}{d}$$

Because the capacitors are in parallel, C_{eq} is the sum of C_1 and C_2:

$$C_{eq} = C_1 + C_2 = \frac{\kappa \epsilon_0 \, ax}{d} + \frac{\epsilon_0 \, a(a-x)}{d}$$

$$= \frac{\epsilon_0 \, a}{d}\left(\kappa x + a - x\right)$$

$$= \frac{\epsilon_0 \, a}{d}\left[(\kappa - 1)x + a\right]$$

Substitute for C_{eq} in the expression for U and simplify to obtain:

$$U = \boxed{\frac{Q^2 d}{2\,\epsilon_0\, a\left[(\kappa - 1)x + a\right]}}$$

(b) The force exerted by the electric field is given by:

$$F = -\frac{dU}{dx}$$

$$= -\frac{d}{dx}\left[\frac{1}{2\,\epsilon_0}\frac{Q^2 d}{a\left[(\kappa - 1)x + a\right]}\right]$$

$$= -\frac{Q^2 d}{2\,\epsilon_0\, a}\frac{d}{dx}\left\{\left[(\kappa - 1)x + a\right]^{-1}\right\}$$

$$= \boxed{\frac{(\kappa - 1)Q^2 d}{2a\,\epsilon_0\left[(\kappa - 1)x + a\right]^2}}$$

(c) Rewrite the result in (b) to obtain:

$$F = \frac{(\kappa - 1)Q^2\left(\dfrac{a\,\epsilon_0}{d}\right)}{2\left(\dfrac{a\,\epsilon_0}{d}\right)^2\left[(\kappa - 1)x + a\right]^2}$$

$$= \frac{(\kappa - 1)Q^2\left(\dfrac{a\,\epsilon_0}{d}\right)}{2C_{eq}^2}$$

$$= \boxed{\frac{(\kappa - 1)a\,\epsilon_0\, V^2}{2d}}$$

Note that this expression is independent of x.

(d) The force originates from the fringing fields around the edges of the capacitor. The effect of the force is to pull the polarized dielectric into the space between the capacitor plates.

Chapter 25
Electric Current and Direct-Current Circuits

Conceptual Problems

13 • A heater consists of a variable resistor (a resistor whose resistance can be varied) connected across an ideal voltage supply. (An ideal voltage supply is one that has a constant emf and a negligible internal resistance.) To increase the heat output, should you decrease the resistance or increase the resistance? Explain your answer.

Determine the Concept You should decrease the resistance. The heat output is given by $P = V^2/R$. Because the voltage across the resistor is constant, decreasing the resistance will increase P.

23 •• In Figure 25-50, the values of the resistances are related as follows: $R_2 = R_3 = 2R_1$. If power P is delivered to R_1, what is the power delivered to R_2 and R_3?

Determine the Concept The power delivered to a resistor varies with the square of the current in the resistor and is directly proportional to the resistance of the resistor ($P = I^2 R$). The current in either resistor 2 or 3 is half the current in resistor 1, and the resistance of either 2 or 3 is half that of resistor 1. Hence the power delivered to either resistor 2 or 3 is one eighth the power delivered to resistor 1.

Current, Current Density, Drift Speed and the Motion of Charges

31 • A 10-gauge copper wire carries a current equal to 20 A. Assuming copper has one free electron per atom, calculate the drift speed of the free electrons in the wire.

Picture the Problem We can relate the drift velocity of the electrons to the current density using $I = nev_d A$. We can find the number density of charge carriers n using $n = \rho N_A / M$, where ρ is the mass density, N_A Avogadro's number, and M the molar mass. We can find the cross-sectional area of 10-gauge wire in Table 25-2.

Use the relation between current and drift velocity to relate I and n:

$$I = nev_d A \Rightarrow v_d = \frac{I}{neA}$$

The number density of charge carriers n is related to the mass density ρ, Avogadro's number N_A, and the molar mass M:

$$n = \frac{\rho N_A}{M}$$

For copper, $\rho = 8.93$ g/cm^3 and $M = 63.55$ g/mol. Substitute and evaluate n:

$$n = \frac{(8.93 \, \text{g/cm}^3)(6.022 \times 10^{23} \, \text{atoms/mol})}{63.55 \, \text{g/mol}}$$

$$= 8.459 \times 10^{28} \, \text{atoms/m}^3$$

Using Table 25-2, find the cross-sectional area of 10-gauge wire:

$$A = 5.261 \, \text{mm}^2$$

Substitute numerical values and evaluate v_d:

$$v_d = \frac{20 \, \text{A}}{(8.459 \times 10^{28} \, \text{m}^{-3})(1.602 \times 10^{-19} \, \text{C})(5.261 \, \text{mm}^2)} = \boxed{0.28 \, \text{mm/s}}$$

33 ••   A length of 10-gauge copper wire and a length of 14-gauge copper wire are welded together end to end. The wires carry a current of 15 A. (*a*) If there is one free electron for each copper atom in each wire, find the drift speed of the electrons in each wire. (*b*) What is the ratio of the magnitude of the current density in the length of10-gauge wire to the magnitude of the current density in the length of 14-gauge wire?

Picture the Problem (*a*) The current will be the same in the two wires and we can relate the drift velocity of the electrons in each wire to their current densities and the cross-sectional areas of the wires. We can find the number density of charge carriers n using $n = \rho N_A / M$, where ρ is the mass density, N_A Avogadro's number, and M the molar mass. We can find the cross-sectional area of 10- and 14-gauge wires in Table 25-2. In Part (*b*) we can use the definition of current density to find the ratio of the magnitudes of the current densities in the 10-gauge and 14-gauge wires.

(*a*) Relate the current density to the drift velocity of the electrons in the 10-gauge wire:

$$\frac{I_{10 \, \text{gauge}}}{A_{10 \, \text{gauge}}} = nev_{d,10} \Rightarrow v_{d,10} = \frac{I_{10 \, \text{gauge}}}{neA_{10 \, \text{gauge}}}$$

The number density of charge carriers n is related to the mass density ρ, Avogadro's number N_A, and the molar mass M:

$$n = \frac{\rho N_A}{M}$$

For copper, $\rho = 8.93$ g/cm^3 and $M = 63.55$ g/mol. Substitute numerical values and evaluate n:

$$n = \frac{\left(8.93\dfrac{\text{g}}{\text{cm}^3}\right)\left(6.022\times10^{23}\dfrac{\text{atoms}}{\text{mol}}\right)}{63.55\dfrac{\text{g}}{\text{mol}}}$$

$$= 8.462\times10^{28}\frac{\text{atoms}}{\text{m}^3}$$

Use Table 25-2 to find the cross-sectional area of 10-gauge wire:

$$A_{10} = 5.261\,\text{mm}^2$$

Substitute numerical values and evaluate $v_{d.10}$:

$$v_{d,10} = \frac{15\,\text{A}}{\left(8.462\times10^{28}\,\text{m}^{-3}\right)\left(1.602\times10^{-19}\,\text{C}\right)\left(5.261\,\text{mm}^2\right)} = 0.210\,\text{mm/s} = \boxed{0.21\,\text{mm/s}}$$

Express the continuity of the current in the two wires:

$$I_{10\,\text{gauge}} = I_{14\,\text{gauge}}$$

or

$$nev_{d,10}A_{10\,\text{gauge}} = nev_{d,14}A_{14\,\text{gauge}}$$

Solve for $v_{d,14}$ to obtain:

$$v_{d,14} = v_{d,10}\frac{A_{10\,\text{gauge}}}{A_{14\,\text{gauge}}}$$

Use Table 25-2 to find the cross-sectional area of 14-gauge wire:

$$A_{14} = 2.081\,\text{mm}^2$$

Substitute numerical values and evaluate $v_{d,14}$:

$$v_{d,14} = \left(0.210\,\text{mm/s}\right)\frac{5.261\,\text{mm}^2}{2.081\,\text{mm}^2}$$

$$= \boxed{0.53\,\text{mm/s}}$$

(b) The ratio of current density, 10-gauge wire to 14-gauge wire, is given by:

$$\frac{J_{10}}{J_{14}} = \frac{\dfrac{I_{10}}{A_{10}}}{\dfrac{I_{14}}{A_{14}}} = \frac{I_{10}A_{14}}{I_{14}A_{10}}$$

Because $I_{10} = I_{14}$:

$$\frac{J_{10}}{J_{14}} = \frac{A_{14}}{A_{10}}$$

Substitute numerical values and evaluate $\dfrac{J_{10}}{J_{14}}$:

$$\frac{J_{10}}{J_{14}} = \frac{2.081\,\text{mm}^2}{5.261\,\text{mm}^2} = \boxed{0.396}$$

35 •• In one of the colliding beams of a planned proton *supercollider*, the protons are moving at nearly the speed of light and the beam current is 5.00-mA. The current density is uniformly distributed throughout the beam. (*a*) How many protons are there per meter of length of the beam? (*b*) If the cross-sectional area of the beam is 1.00×10^{-6} m^2, what is the number density of protons? (*c*) What is the magnitude of the current density in this beam?

Picture the Problem We can relate the number of protons per meter N to the number n of free charge-carrying particles per unit volume in a beam of cross-sectional area A and then use the relation between current and drift velocity to relate n to I.

(*a*) Express the number of protons per meter N in terms of the number n of free charge-carrying particles per unit volume in a beam of cross-sectional area A:

$$N = nA \qquad (1)$$

Use the relation between current and drift velocity to relate I and n:

$$I = enAv \Rightarrow n = \frac{I}{eAv}$$

Substitute for n and simplify to obtain:

$$N = \frac{IA}{eAv} = \frac{I}{ev}$$

Substitute numerical values and evaluate N:

$$N = \frac{5.00\,\text{mA}}{\left(1.602 \times 10^{-19}\,\text{C}\right)\left(2.998 \times 10^8\,\text{m/s}\right)}$$

$$= 1.041 \times 10^8\,\text{m}^{-1}$$

$$= \boxed{1.04 \times 10^8\,\text{m}^{-1}}$$

(*b*) From equation (1) we have:

$$n = \frac{N}{A} = \frac{1.041 \times 10^8\,\text{m}^{-1}}{1.00 \times 10^{-6}\,\text{m}^2}$$

$$= \boxed{1.04 \times 10^{14}\,\text{m}^{-3}}$$

(*c*) The magnitude of the current density in this beam is given by:

$$J = \frac{I}{A} = \frac{5.00\,\text{mA}}{1.00 \times 10^{-6}\,\text{m}^2} = \boxed{5.00\,\text{kA/m}^2}$$

Resistance, Resistivity and Ohm's Law

39 •   A potential difference of 100 V across the terminals of a resistor produces a current of 3.00 A in the resistor. (*a*) What is the resistance of the resistor? (*b*) What is the current in the resistor when the potential difference is only 25.0 V? (Assume the resistance of the resistor remains constant.)

Picture the Problem We can apply Ohm's law to both parts of this problem, solving first for R and then for I.

(*a*) Apply Ohm's law to obtain:

$$R = \frac{V}{I} = \frac{100\,\text{V}}{3.00\,\text{A}} = \boxed{33.3\,\Omega}$$

(*b*) Apply Ohm's law a second time to obtain:

$$I = \frac{V}{R} = \frac{25.0\,\text{V}}{33.3\,\Omega} = \boxed{0.750\,\text{A}}$$

41 •   An extension cord consists of a pair of 30-m-long 16-gauge copper wires. What is the potential difference that must be applied across one of the wires if it is to carry a current of 5.0 A?

Picture the Problem We can use Ohm's law in conjunction with $R = \rho L / A$ to find the potential difference across one wire of the extension cord.

Using Ohm's law, express the potential difference across one wire of the extension cord:

$$V = IR$$

Relate the resistance of the wire to its resistivity ρ, cross-sectional area A, and length L:

$$R = \rho \frac{L}{A}$$

Substitute for R to obtain:

$$V = \rho \frac{LI}{A}$$

Substitute numerical values (see Table 25-1 for the resistivity of copper and Table 25-2 for the cross-sectional area of 16-gauge wire) and evaluate V:

$$V = \left(1.7 \times 10^{-8}\,\Omega \cdot \text{m}\right) \frac{(30\,\text{m})(5.0\,\text{A})}{1.309\,\text{mm}^2}$$

$$= \boxed{1.9\,\text{V}}$$

45 ••   A 1.00-m-long wire has a resistance equal to 0.300 Ω. A second wire made of identical material has a length of 2.00 m and a mass equal to the mass of the first wire. What is the resistance of the second wire?

Picture the Problem We can use $R = \rho L/A$ to relate the resistance of the wires to their lengths, resistivities, and cross-sectional areas. To find the resistance of the second wire, we can use the fact that the volumes of the two wires are the same to relate the cross-sectional area of the first wire to the cross-sectional area of the second wire.

Relate the resistance of the first wire to its resistivity, cross-sectional area, and length:

$$R = \rho\frac{L}{A}$$

Relate the resistance of the second wire to its resistivity, cross-sectional area, and length:

$$R' = \rho\frac{L'}{A'}$$

Divide the second of these equations by the first to obtain:

$$\frac{R'}{R} = \frac{\rho\dfrac{L'}{A'}}{\rho\dfrac{L}{A}} = \frac{L'}{L}\frac{A}{A'} \Rightarrow R' = 2\frac{A}{A'}R \quad (1)$$

Express the relationship between the volume V of the first wire and the volume V' of the second wire:

$$V = V' \text{ or } LA = L'A' \Rightarrow \frac{A}{A'} = \frac{L'}{L} = 2$$

Substituting for $\dfrac{A}{A'}$ in equation (1) yields:

$$R' = 2(2)R = 4R$$

Because $R = 3.00\ \Omega$:

$$R' = 4(0.300\,\Omega) = \boxed{1.20\,\Omega}$$

49 •••   Consider a wire of length L in the shape of a truncated cone. The radius of the wire varies with distance x from the narrow end according to $r = a + [(b - a)/L]x$, where $0 < x < L$. Derive an expression for the resistance of this wire in terms of its length L, radius a, radius b and resistivity ρ. *Hint: Model the wire as a series combination of a large number of thin disks. Assume the current is uniformly distributed on a cross section of the cone.*

Picture the Problem The element of resistance we use is a segment of length dx and cross-sectional area $\pi[a + (b - a)x/L]^2$. Because these resistance elements are in series, integrating over them will yield the resistance of the wire.

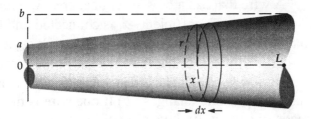

Express the resistance of the chosen element of resistance:

$$dR = \rho \frac{dx}{A} = \frac{\rho}{\pi[a+(b-a)(x/L)]^2}dx$$

Integrate dR from $x = 0$ to $x = L$ and simplify to obtain:

$$R = \frac{\rho}{\pi}\int_0^L \frac{dx}{[a+(b-a)(x/L)]^2}$$

$$= \frac{\rho L}{\pi(b-a)}\left(\frac{1}{a} - \frac{1}{a+(b-a)}\right)$$

$$= \boxed{\frac{\rho L}{\pi ab}}$$

Temperature Dependence of Resistance

53 • At what temperature will the resistance of a copper wire be 10 percent greater than its resistance at 20°C?

Picture the Problem The resistance of the copper wire increases with temperature according to $R_{t_C} = R_{20}[1+\alpha(t_C - 20C°)]$. We can replace R_{t_C} by $1.1R_{20}$ and solve for t_C to find the temperature at which the resistance of the wire will be 110% of its value at 20°C.

Express the resistance of the wire at $1.10R_{20}$:

$$1.10R_{20} = R_{20}[1+\alpha(t_C - 20C°)]$$

Simplifying this expression yields:

$$0.10 = \alpha(t_C - 20C°)$$

Solve to t_C to obtain:

$$t_C = \frac{0.10}{\alpha} + 20C°$$

Substitute numerical values (see Table 25-1 for the temperature coefficient of resistivity of copper) and evaluate t_C:

$$t_C = \frac{0.10}{3.9\times10^{-3}\text{ K}^{-1}} + 20C° = \boxed{46°C}$$

57 •••   A wire that has a cross-sectional area A, a length L_1, a
resistivity ρ_1, and a temperature coefficient α_1 is connected end to end to a
second wire that has the same cross-sectional area, a length L_2, a resistivity ρ_2,
and a temperature coefficient α_2, so that the wires carry the same current.
(*a*) Show that if $\rho_1 L_1 \alpha_1 + \rho_2 L_2 \alpha_2 = 0$, then the total resistance is independent of
temperature for small temperature changes. (*b*) If one wire is made of carbon and
the other wire is made of copper, find the ratio of their lengths for which the total
resistance is approximately independent of temperature.

Picture the Problem Expressing the total resistance of the two current-carrying
(and hence warming) wires connected in series in terms of their resistivities,
temperature coefficients of resistivity, lengths and temperature change will lead
us to an expression in which, if $\rho_1 L_1 \alpha_1 + \rho_2 L_2 \alpha_2 = 0$, the total resistance is
temperature independent. In Part (*b*) we can apply the condition that

$$\rho_1 L_1 \alpha_1 + \rho_2 L_2 \alpha_2 = 0$$

to find the ratio of the lengths of the carbon and copper wires.

(*a*) Express the total resistance of these two wires connected in series:

$$R = R_1 + R_2 = \rho_1 \frac{L_1}{A}\left(1 + \alpha_1 \Delta T\right) + \rho_2 \frac{L_2}{A}\left(1 + \alpha_2 \Delta T\right)$$

$$= \frac{1}{A}\left[\rho_1 L_1 \left(1 + \alpha_1 \Delta T\right) + \rho_2 L_2 \left(1 + \alpha_2 \Delta T\right)\right]$$

Expand and simplify this expression to obtain:

$$R = \frac{1}{A}\left[\rho_1 L_1 + \rho_2 L_2 + \left(\rho_1 L_1 \alpha_1 + \rho_1 L_1 \alpha_2\right)\Delta T\right]$$

If $\rho_1 L_1 \alpha_1 + \rho_2 L_2 \alpha_2 = 0$, then:

$$R = \boxed{\frac{1}{A}\left[\rho_1 L_1 + \rho_2 L_2\right]}$$ independently of

the temperature.

(*b*) Apply the condition for $\rho_C L_C \alpha_C + \rho_{Cu} L_{Cu} \alpha_{Cu} = 0$
temperature independence obtained
in (*a*) to the carbon and copper
wires:

Solve for the ratio of L_{Cu} to L_C: $$\frac{L_{Cu}}{L_C} = -\frac{\rho_C \alpha_C}{\rho_{Cu} \alpha_{Cu}}$$

Substitute numerical values (see Table 25-1 for the temperature coefficient of resistivity of carbon and copper) and evaluate the ratio of L_{Cu} to L_C:

$$\frac{L_{Cu}}{L_C} = -\frac{\left(3500\times10^{-8}\,\Omega\cdot m\right)\left(-0.5\times10^{-3}\,K^{-1}\right)}{\left(1.7\times10^{-8}\,\Omega\cdot m\right)\left(3.9\times10^{-3}\,K^{-1}\right)} \approx \boxed{3\times10^2}$$

Energy in Electric Circuits

63 • (*a*) How much power is delivered by the battery in Problem 62 due to the chemical reactions within the battery when the current in the battery is 20 A? (*b*) How much of this power is delivered to the starter when the current in the battery is 20 A? (*c*) By how much does the chemical energy of the battery decrease if the current in the starter is 20 A for 7.0 s? (*d*) How much energy is dissipated in the battery during these 7.0 seconds?

Picture the Problem We can find the power delivered by the battery from the product of its emf and the current it delivers. The power delivered to the battery can be found from the product of the potential difference across the terminals of the starter (or across the battery when current is being drawn from it) and the current being delivered to it. In Part (*c*) we can use the definition of power to relate the decrease in the chemical energy of the battery to the power it is delivering and the time during which current is drawn from it. In Part (*d*) we can use conservation of energy to relate the energy delivered by the battery to the heat developed in the battery and the energy delivered to the starter

(*a*) Express the power delivered by the battery as a function of its emf and the current it delivers:

$$P = \mathcal{E}I = (12.0\,V)(20\,A) = 240\,W$$
$$= \boxed{0.24\,kW}$$

(*b*) Relate the power delivered to the starter to the potential difference across its terminals:

$$P_{starter} = V_{starter}I = (11.4\,V)(20\,A)$$
$$= 228\,W = \boxed{0.23\,kW}$$

(*c*) Use the definition of power to express the decrease in the chemical energy of the battery as it delivers current to the starter:

$$\Delta E = P\Delta t$$
$$= (240\,W)(7.0\,s) = 1680\,J$$
$$= \boxed{1.7\,kJ}$$

(d) Use conservation of energy to relate the energy delivered by the battery to the heat developed in the battery and the energy delivered to the starter:

$$E_{\substack{delivered \\ by\ battery}} = E_{\substack{transformed \\ into\ heat}} + E_{\substack{delivered \\ to\ starter}}$$
$$= Q + E_{\substack{delivered \\ to\ starter}}$$

Express the energy delivered by the battery and the energy delivered to the starter in terms of the rate at which this energy is delivered:

$$P\Delta t = Q + P_s \Delta t \Rightarrow Q = (P - P_s)\Delta t$$

Substitute numerical values and evaluate Q:

$$Q = (240\,\text{W} - 228\,\text{W})(7.0\,\text{s}) = \boxed{84\,\text{J}}$$

67 •• A lightweight electric car is powered by a series combination of ten 12.0-V batteries, each having negligible internal resistance. Each battery can deliver a charge of 160 A·h before needing to be recharged. At a speed of 80.0 km/h, the average force due to air drag and rolling friction is 1.20 kN. (a) What must be the minimum power delivered by the electric motor if the car is to travel at a speed of 80.0 km/h? (b) What is the total charge, in coulombs, that can be delivered by the series combination of ten batteries before recharging is required? (c) What is the total electrical energy delivered by the ten batteries before recharging? (d) How far can the car travel (at 80.0 km/h) before the batteries must be recharged? (e) What is the cost per kilometer if the cost of recharging the batteries is 9.00 cents per kilowatt-hour?

Picture the Problem We can use $P = fv$ to find the power the electric motor must develop to move the car at 80 km/h against a frictional force of 1200 N. We can find the total charge that can be delivered by the 10 batteries using $\Delta Q = NI\Delta t$. The total electrical energy delivered by the 10 batteries before recharging can be found using the definition of emf. We can find the distance the car can travel from the definition of work and the cost per kilometer of driving the car this distance by dividing the cost of the required energy by the distance the car has traveled.

(a) Express the power the electric motor must develop in terms of the speed of the car and the friction force:

$$P = fv = (1.20\,\text{kN})(80.0\,\text{km/h})$$
$$= \boxed{26.7\,\text{kW}}$$

(b) Because the batteries are in series, the total charge that can be delivered before charging is the same as the charge from a single battery:

$$\Delta Q = I\Delta t = (160\,\text{A}\cdot\text{h})\left(\frac{3600\,\text{s}}{\text{h}}\right)$$

$$= \boxed{576\,\text{kC}}$$

(c) Use the definition of emf to express the total electrical energy available in the batteries:

$$W = Q\varepsilon = 10(576\,\text{kC})(12.0\,\text{V})$$

$$= 69.12\,\text{MJ} = \boxed{69.1\,\text{MJ}}$$

(d) Relate the amount of work the batteries can do to the work required to overcome friction:

$$W = fd \Rightarrow d = \frac{W}{f}$$

Substitute numerical values and evaluate d:

$$d = \frac{69.12\,\text{MJ}}{1.20\,\text{kN}} = \boxed{57.6\,\text{km}}$$

(e) The cost per kilometer is the ratio of the cost of the energy to the distance traveled before recharging:

$$\text{Cost/km} = \frac{\left(\dfrac{\$0.0900}{\text{kW}\cdot\text{h}}\right)\varepsilon I t}{d}$$

Substitute numerical values and calculate the cost per kilometer:

$$\text{Cost/km} = \frac{\left(\dfrac{\$0.0900}{\text{kW}\cdot\text{h}}\right)(120\,\text{V})(160\,\text{A}\cdot\text{h})}{5.76\,\text{km}} = \boxed{\$0.300/\text{km}}$$

Combinations of Resistors

69 •   If the potential drop from point a to point b (Figure 25-52) is 12.0 V, find the current in each resistor.

Picture the Problem We can apply Ohm's law to find the current through each resistor.

Apply Ohm's law to each of the resistors to find the current flowing through each:

$$I_4 = \frac{V}{4.00\,\Omega} = \frac{12.0\,\text{V}}{4.00\,\Omega} = \boxed{3.00\,\text{A}}$$

$$I_3 = \frac{V}{3.00\,\Omega} = \frac{12.0\,\text{V}}{3.00\,\Omega} = \boxed{4.00\,\text{A}}$$

and

$$I_6 = \frac{V}{6.00\,\Omega} = \frac{12.0\,\text{V}}{6.00\,\Omega} = \boxed{2.00\,\text{A}}$$

Remarks: You would find it instructive to use Kirchhoff's junction rule (conservation of charge) to confirm our values for the currents through the three resistors.

73 •• A 5.00-V power supply has an internal resistance of 50.0 Ω. What is the smallest resistor that can be put in series with the power supply so that the voltage drop across the resistor is larger than 4.50 V?

Picture the Problem Let r represent the resistance of the internal resistance of the power supply, $\mathcal{E}$ the emf of the power supply, R the resistance of the external resistor to be placed in series with the power supply, and I the current drawn from the power supply. We can use Ohm's law to express the potential difference across R and apply Kirchhoff's loop rule to express the current through R in terms of $\mathcal{E}$, r, and R.

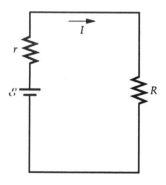

Express the potential difference across the resistor whose resistance is R:

$$V_R = IR \qquad\qquad (1)$$

Apply Kirchhoff's loop rule to the circuit to obtain:

$$\mathcal{E} - Ir - IR = 0 \Rightarrow I = \frac{\mathcal{E}}{r+R}$$

Substitute in equation (1) to obtain:

$$V_R = \left(\frac{\mathcal{E}}{r+R}\right)R \Rightarrow R = \frac{V_R r}{\mathcal{E} - V_R}$$

Substitute numerical values and evaluate R:

$$R = \frac{(4.50\,\text{V})(50.0\,\Omega)}{5.00\,\text{V} - 4.50\,\text{V}} = \boxed{0.45\,\text{k}\Omega}$$

77 •• A length of wire has a resistance of 120 Ω. The wire is cut into pieces that have the same length, and then the wires are connected in parallel. The resistance of the parallel arrangement is 1.88 Ω. Find the number of pieces into which the wire was cut.

Picture the Problem We can use the equation for N identical resistors connected in parallel to relate N to the resistance R of each piece of wire and the equivalent resistance R_{eq}.

Express the resistance of the N pieces connected in parallel:

$$\frac{1}{R_{eq}} = \frac{N}{R}$$

where R is the resistance of one of the N pieces.

Relate the resistance of one of the N pieces to the resistance of the wire:

$$R = \frac{R_{wire}}{N}$$

Substitute for R to obtain:

$$\frac{1}{R_{eq}} = \frac{N^2}{R_{wire}} \Rightarrow N = \sqrt{\frac{R_{wire}}{R_{eq}}}$$

Substitute numerical values and evaluate N:

$$N = \sqrt{\frac{120\,\Omega}{1.88\,\Omega}} = \boxed{8\ \text{pieces}}$$

Kirchhoff's Rules

81 • In Figure 25-59, the battery's emf is 6.00 V and R is 0.500 Ω. The rate of Joule heating in R is 8.00 W. (*a*) What is the current in the circuit? (*b*) What is the potential difference across R? (*c*) What is the resistance r?

Picture the Problem We can relate the current provided by the source to the rate of Joule heating using $P = I^2 R$ and use Ohm's law and Kirchhoff's rules to find the potential difference across R and the value of r.

(*a*) Relate the current I in the circuit to rate at which energy is being dissipated in the form of Joule heat:

$$P = I^2 R \Rightarrow I = \sqrt{\frac{P}{R}}$$

Substitute numerical values and evaluate I:

$$I = \sqrt{\frac{8.00\,\text{W}}{0.500\,\Omega}} = \boxed{4.00\,\text{A}}$$

(b) Apply Ohm's law to find V_R:

$$V_R = IR = (4.00\,\text{A})(0.500\,\Omega) = \boxed{2.00\,\text{V}}$$

(c) Apply Kirchhoff's loop rule to obtain:

$$\mathcal{E} - Ir - IR = 0 \Rightarrow r = \frac{\mathcal{E} - IR}{I} = \frac{\mathcal{E}}{I} - R$$

Substitute numerical values and evaluate r:

$$r = \frac{6.00\,\text{V}}{4.00\,\text{A}} - 0.500\,\Omega = \boxed{1.00\,\Omega}$$

85 •• In the circuit shown in Figure 25-62, the batteries have negligible internal resistance. Find (a) the current in each branch of the circuit, (b) the potential difference between point a and point b, and (c) the power supplied by each battery.

Picture the Problem Let I_1 be the current delivered by the left battery, I_2 the current delivered by the right battery, and I_3 the current through the 6.00-Ω resistor, directed down. We can apply Kirchhoff's rules to obtain three equations that we can solve simultaneously for I_1, I_2, and I_3. Knowing the currents in each branch, we can use Ohm's law to find the potential difference between points a and b and the power delivered by both the sources.

(a) Apply Kirchhoff's junction rule at junction a:

$$I_{4\Omega} + I_{3\Omega} = I_{6\Omega} \qquad (1)$$

Apply Kirchhoff's loop rule to a loop around the outside of the circuit to obtain:

$$12.0\,\text{V} - (4.00\,\Omega)I_{4\Omega} + (3.00\,\Omega)I_{3\Omega}$$
$$-12.0\,\text{V} = 0$$

or

$$-(4.00\,\Omega)I_{4\Omega} + (3.00\,\Omega)I_{3\Omega} = 0 \quad (2)$$

Apply Kirchhoff's loop rule to a loop around the left-hand branch of the circuit to obtain:

$$12.0\,\text{V} - (4.00\,\Omega)I_{4\Omega}$$
$$-(6.00\,\Omega)I_{6\Omega} = 0 \qquad (3)$$

Solving equations (1), (2), and (3) simultaneously yields:

$$I_{4\Omega} = \boxed{0.667\,\text{A}}, \ I_{3\Omega} = \boxed{0.889\,\text{A}},$$

and

$$I_{6\Omega} = \boxed{1.56\,\text{A}}$$

(b) Apply Ohm's law to find the potential difference between points a and b:

$$V_{ab} = (6.00\,\Omega)I_{6\Omega} = (6.00\,\Omega)(1.56\,\text{A})$$
$$= \boxed{9.36\,\text{V}}$$

(c) Express the power delivered by the 12.0-V battery in the left-hand branch of the circuit:

$$P_{\text{left}} = \mathcal{E} I_{4\Omega}$$
$$= (12.0\,\text{V})(0.667\,\text{A}) = \boxed{8.00\,\text{W}}$$

Express the power delivered by the 12.0-V battery in the right-hand branch of the circuit:

$$P_{\text{right}} = \mathcal{E} I_{3\Omega}$$
$$= (12.0\,\text{V})(0.889\,\text{A}) = \boxed{10.7\,\text{W}}$$

89 •••   For the circuit shown in Figure 25-65, find the potential difference between point a and point b.

Picture the Problem Let I_1 be the current in the left branch resistor, directed up; let I_3 be the current, directed down, in the middle branch; and let I_2 be the current in the right branch, directed up. We can apply Kirchhoff's rules to find I_3 and then the potential difference between points a and b.

Relate the potential at a to the potential at b:	$V_a - R_4 I_3 - 4\,\text{V} = V_b$ or $V_a - V_b = R_4 I_3 + 4\,\text{V}$
Apply Kirchhoff's junction rule at a to obtain:	$I_1 + I_2 = I_3$ (1)
Apply the loop rule to a loop around the outside of the circuit to obtain:	$2.00\,\text{V} - (1.00\,\Omega) I_1 + (1.00\,\Omega) I_2$ $-2.00\,\text{V} + (1.00\,\Omega) I_2 - (1.00\,\Omega) I_1 = 0$ or $I_1 - I_2 = 0$ (2)
Apply the loop rule to the left side of the circuit to obtain:	$2.00\,\text{V} - (1.00\,\Omega) I_1 - (4.00\,\Omega) I_3$ $-4.00\,\text{V} - (1.00\,\Omega) I_1 = 0$ or $-(1.00\,\Omega) I_1 - (2.00\,\Omega) I_3 = 1.00\,\text{V}$ (3)
Solve equations (1), (2), and (3) simultaneously to obtain:	$I_1 = -0.200\,\text{A}$, $I_2 = -0.200\,\text{A}$, and $I_3 = -0.400\,\text{A}$ where the minus signs indicate that the currents flow in opposite directions to the directions chosen.

Substitute to obtain:

$$V_a - V_b = (4.00\,\Omega)(-0.400\,\text{A}) + 4.00\,\text{V}$$

$$= \boxed{2.40\,\text{V}}$$

Remarks: Note that point a is at the higher potential.

Ammeters and Voltmeters

91 ••   The voltmeter shown in Figure 25-67 can be modeled as an ideal voltmeter (a voltmeter that has an infinite internal resistance) in parallel with a 10.0 MΩ resistor. Calculate the reading on the voltmeter when (a) $R = 1.00$ kΩ, (b) $R = 10.0$ kΩ, (c) $R = 1.00$ MΩ, (d) $R = 10.0$ MΩ, and (e) $R = 100$ MΩ. (f) What is the largest value of R possible if the measured voltage is to be within 10 percent of the *true* voltage (that is, the voltage drop across R without the voltmeter in place)?

Picture the Problem Let I be the current drawn from source and R_{eq} the resistance equivalent to R and 10 MΩ connected in parallel and apply Kirchhoff's loop rule to express the measured voltage V across R as a function of R.

The voltage measured by the voltmeter is given by:	$V = IR_{eq}$	(1)

Apply Kirchhoff's loop rule to the circuit to obtain:

$$10.0\,\text{V} - IR_{eq} - I(2R) = 0$$

Solving for I yields:

$$I = \frac{10.0\,\text{V}}{R_{eq} + 2R}$$

Express R_{eq} in terms of R and 10.0-MΩ resistance in parallel with it:

$$\frac{1}{R_{eq}} = \frac{1}{10.0\,\text{M}\Omega} + \frac{1}{R}$$

Solving for R_{eq} yields:

$$R_{eq} = \frac{(10.0\,\text{M}\Omega)R}{R + 10.0\,\text{M}\Omega}$$

Substitute for I in equation (1) and simplify to obtain:

$$V = \left(\frac{10.0\,\text{V}}{R_{eq} + 2R}\right)R_{eq} = \frac{10.0\,\text{V}}{1 + \dfrac{2R}{R_{eq}}}$$

Substitute for R_{eq} and simplify to obtain:

$$V = \frac{(10.0\,\text{V})(5.0\,\text{M}\Omega)}{R + 15.0\,\text{M}\Omega} \qquad (2)$$

(a) Evaluate equation (2) for
$R = 1.00$ kΩ:

$$V = \frac{(10.0\,\text{V})(5.0\,\text{M}\Omega)}{1.00\,\text{k}\Omega + 15.0\,\text{M}\Omega} = \boxed{3.3\,\text{V}}$$

(b) Evaluate equation (2) for
$R = 10.0$ kΩ:

$$V = \frac{(10.0\,\text{V})(5.0\,\text{M}\Omega)}{10.0\,\text{k}\Omega + 15.0\,\text{M}\Omega} = \boxed{3.3\,\text{V}}$$

(c) Evaluate equation (2) for
$R = 1.00$ MΩ:

$$V = \frac{(10.0\,\text{V})(5.0\,\text{M}\Omega)}{1.00\,\text{M}\Omega + 15.0\,\text{M}\Omega} = \boxed{3.1\,\text{V}}$$

(d) Evaluate equation (2) for
$R = 10.0$ MΩ:

$$V = \frac{(10.0\,\text{V})(5.0\,\text{M}\Omega)}{10.0\,\text{M}\Omega + 15.0\,\text{M}\Omega} = \boxed{2.0\,\text{V}}$$

(e) Evaluate equation (2) for
$R = 100$ MΩ:

$$V = \frac{(10.0\,\text{V})(5.0\,\text{M}\Omega)}{100\,\text{M}\Omega + 15.0\,\text{M}\Omega} = \boxed{0.43\,\text{V}}$$

(f) Express the condition that the measured voltage to be within 10 percent of the *true* voltage V_{true}:

$$\frac{V_{\text{true}} - V}{V_{\text{true}}} = 1 - \frac{V}{V_{\text{true}}} < 0.1$$

Substitute for V and V_{true} to obtain:

$$1 - \frac{\dfrac{(10.0\,\text{V})(5.0\,\text{M}\Omega)}{R + 15.0\,\text{M}\Omega}}{IR} < 0.1$$

Because $I = 10.0$ V/$3R$:

$$1 - \frac{\dfrac{(10.0\,\text{V})(5.0\,\text{M}\Omega)}{R + 15.0\,\text{M}\Omega}}{\dfrac{10.0}{3}\,\text{V}} < 0.1$$

Solving for R yields:

$$R < \frac{1.5\,\text{M}\Omega}{0.90} = \boxed{1.67\,\text{M}\Omega}$$

RC Circuits

97 •• In the circuit in Figure 25-69, the emf equals 50.0 V and the capacitance equals 2.00 μF. Switch S is opened after having been closed for a long time, and 4.00 s later the voltage drop across the resistor is 20.0 V. Find the resistance of the resistor.

Picture the Problem We can find the resistance of the circuit from its time constant and use Ohm's law and the expression for the current in a charging RC circuit to express τ as a function of time, V_0, and $V(t)$.

Express the resistance of the resistor in terms of the time constant of the circuit:	$$R = \frac{\tau}{C} \qquad (1)$$
Using Ohm's law, express the voltage drop across the resistor as a function of time:	$$V(t) = I(t)R$$
Express the current in the circuit as a function of the elapsed time after the switch is closed:	$$I(t) = I_0 e^{-t/\tau}$$
Substitute for $I(t)$ to obtain:	$$V(t) = I_0 e^{-t/\tau} R = (I_0 R) e^{-t/\tau} = V_0 e^{-t/\tau}$$
Take the natural logarithm of both sides of the equation and solve for τ to obtain:	$$\tau = -\frac{t}{\ln\left[\dfrac{V(t)}{V_0}\right]}$$
Substitute for τ in equation (1) to obtain:	$$R = -\frac{t}{C \ln\left[\dfrac{V(t)}{V_0}\right]}$$
Substitute numerical values and evaluate R using the data given for $t = 4.00$ s:	$$R = -\frac{4.00\,\text{s}}{(2.00\,\mu\text{F})\ln\left(\dfrac{20.0\,\text{V}}{50.0\,\text{V}}\right)}$$ $$= \boxed{2.18\,\text{M}\Omega}$$

105 •••   In the circuit shown in Figure 25-74, the capacitor has a capacitance of 2.50 μF and the resistor has a resistance of 0.500 MΩ. Before the switch is closed, the potential drop across the capacitor is 12.0 V, as shown. Switch S is closed at $t = 0$. (*a*) What is the current immediately after switch S is closed? (*b*) At what time t is the voltage across the capacitor 24.0 V?

Picture the Problem We can apply Kirchhoff's loop rule to the circuit immediately after the switch is closed in order to find the initial current I_0. We can find the time at which the voltage across the capacitor is 24.0 V by again applying Kirchhoff's loop rule to find the voltage across the resistor when this condition is satisfied and then using the expression $I(t) = I_0 e^{-t/\tau}$ for the current through the resistor as a function of time and solving for t.

(a) Apply Kirchhoff's loop rule to the circuit immediately after the switch is closed:

$$\mathcal{E} - 12.0\,\text{V} - I_0 R = 0$$

Solving for I_0 yields:

$$I_0 = \frac{\mathcal{E} - 12.0\,\text{V}}{R}$$

Substitute numerical values and evaluate I_0:

$$I_0 = \frac{36.0\,\text{V} - 12.0\,\text{V}}{0.500\,\text{M}\Omega} = \boxed{48.0\,\mu\text{A}}$$

(b) Apply Kirchhoff's loop rule to the circuit when $V_C = 24.0$ V and solve for V_R:

$$36.0\,\text{V} - 24.0\,\text{V} - I(t)R = 0$$
and
$$I(t)R = 12.0\,\text{V}$$

Express the current through the resistor as a function of I_0 and τ:

$$I(t) = I_0 e^{-t/\tau} \text{ where } \tau = RC.$$

Substitute to obtain:

$$RI_0 e^{-t/\tau} = 12.0\,\text{V} \Rightarrow e^{-t/\tau} = \frac{12.0\,\text{V}}{RI_0}$$

Take the natural logarithm of both sides of the equation to obtain:

$$-\frac{t}{\tau} = \ln\left(\frac{12.0\,\text{V}}{RI_0}\right)$$

Solving for t yields:

$$t = -\tau \ln\left(\frac{12.0\,\text{V}}{RI_0}\right) = -RC \ln\left(\frac{12.0\,\text{V}}{RI_0}\right)$$

Substitute numerical values and evaluate t:

$$t = -(0.500\,\text{M}\Omega)(2.50\,\mu\text{F})\ln\left[\frac{12.0\,\text{V}}{(0.500\,\text{M}\Omega)(48.0\,\mu\text{A})}\right] = \boxed{0.866\text{s}}$$

General Problems

107 ••   In Figure 25-75, $R_1 = 4.00\ \Omega$, $R_2 = 6.00\ \Omega$, $R_3 = 12.0\ \Omega$, and the battery emf is 12.0 V. Denote the currents through these resistors as I_1, I_2 and I_3, respectively, (a) Decide which of the following inequalities holds for this circuit. Explain your answer conceptually. (1) $I_1 > I_2 > I_3$, (2) $I_2 = I_3$, (3) $I_3 > I_2$, (4) None of the above (b) To verify that your answer to Part (a) is correct, calculate all three currents.

Determine the Concept We can use Kirchhoff's rules in Part (*b*) to confirm our choices in Part (*a*).

(*a*) 1. The potential drops across R_2 and R_3 are equal, so $I_2 > I_3$. The current in R_1 equals the sum of the currents in I_2 and I_3, so I_1 is greater than either I_2 or I_3.

(*b*) Apply Kirchhoff's junction rule to obtain:

$$I_1 - I_2 - I_3 = 0 \qquad (1)$$

Applying Kirchhoff's loop rule in the clockwise direction to the loop defined by the two resistors in parallel yields:

$$R_3 I_3 - R_2 I_2 = 0$$

or

$$0 I_1 - R_2 I_2 + R_3 I_3 = 0 \qquad (2)$$

Apply Kirchhoff's loop rule in the clockwise direction to the loop around the perimeter of the circuit to obtain:

$$\mathcal{E} - R_1 I_1 - R_2 I_2 = 0$$

or

$$R_1 I_1 + R_2 I_2 - 0 I_3 = \mathcal{E} \qquad (3)$$

Substituting numerical values in equations (1), (2), and (3) yields:

$$I_1 - I_2 - I_3 = 0$$
$$0 I_1 - (6.00\,\Omega) I_2 + (12.0\,\Omega) I_3 = 0$$
$$(4.00\,\Omega) I_1 + (6.00\,\Omega) I_2 - 0 I_3 = 12.0\text{ V}$$

Solve this system of three equations in three unknowns for the currents in the branches of the circuit to obtain:

$$I_1 = \boxed{1.50\text{ A}}, \; I_2 = \boxed{1.00\text{ A}},$$

and $I_3 = \boxed{0.50\text{ A}}$, confirming our choice in Part (*a*).

111 ••   You are running an experiment that uses an accelerator that produces a 3.50-μA proton beam. Each proton in the beam has 60.0-MeV of kinetic energy. The protons impinge upon, and come to rest inside, a 50.0-g copper target within a vacuum chamber. You are concerned that the target will get too hot and melt the solder on some connecting wires that are crucial to the experiment. (*a*) Determine the number of protons that strike the target per second. (*b*) Find the amount of energy delivered to the target each second. (*c*) Determine how much time elapses before the target temperature increases to 300°C? (Neglect any heat released by the target.)

Picture the Problem Knowing the beam current and charge per proton, we can use $I = ne$ to determine the number of protons striking the target per second. The energy deposited per second is the power delivered to the target and is given by $P = IV$. We can find the elapsed time before the target temperature rises 300°C using $\Delta Q = P\Delta t = mc_{Cu}\Delta T$.

(a) Relate the current to the number of protons per second n arriving at the target:

$$I = ne \Rightarrow n = \frac{I}{e}$$

Substitute numerical values and evaluate n:

$$n = \frac{3.50\,\mu\text{A}}{1.602 \times 10^{-19}\,\text{C}} = \boxed{2.18 \times 10^{13}\ \text{s}^{-1}}$$

(b) Express the power of the beam in terms of the beam current and energy:

$$P = IV = (3.50\,\mu\text{A})\left(60.0\frac{\text{MeV}}{\text{proton}}\right)$$

$$= \boxed{210\ \text{W}}$$

(c) Relate the energy delivered to the target to its heat capacity and temperature change:

$$\Delta Q = P\Delta t = C_{\text{Cu}}\Delta T = mc_{\text{Cu}}\Delta T$$

Solving for Δt yields:

$$\Delta t = \frac{mc_{\text{Cu}}\Delta T}{P}$$

Substitute numerical values (see Table 19-1 for the specific heat of copper)and evaluate Δt:

$$\Delta t = \frac{(50.0\,\text{g})(0.386\,\text{kJ/kg}\cdot\text{K})(300^\circ\text{C})}{210\,\text{J/s}}$$

$$= \boxed{27.6\,\text{s}}$$

Chapter 26
The Magnetic Field

Conceptual Problems

1   •   When the axis of a cathode-ray tube is horizontal in a region in which there is a magnetic field that is directed vertically upward, the electrons emitted from the cathode follow one of the dashed paths to the face of the tube in Figure 26-30. The correct path is (*a*) 1, (*b*) 2, (*c*) 3, (*d*) 4, (*e*) 5.

Determine the Concept Because the electrons are initially moving at 90° to the magnetic field, they will be deflected in the direction of the magnetic force acting on them. Use the right-hand rule based on the expression for the magnetic force acting on a moving charge $\vec{F} = q\vec{v} \times \vec{B}$, remembering that, for a negative charge, the force is in the direction opposite that indicated by the right-hand rule, to convince yourself that the particle will follow the path whose terminal point on the screen is 2. $\boxed{(b)}$ is correct.

3   •   A *flicker bulb* is a light bulb that has a long, thin flexible filament. It is meant to be plugged into an ac outlet that delivers current at a frequency of 60 Hz. There is a small permanent magnet inside the bulb. When the bulb is plugged in the filament oscillates back and forth. At what frequency does it oscillate? Explain.

Determine the Concept Because the alternating current running through the filament is changing direction every 1/60 s, the filament experiences a force that changes direction at the frequency of the current.

7   •   In a velocity selector, the speed of the undeflected charged particle is given by the ratio of the magnitude of the electric field to the magnitude of the magnetic field. Show that E/B in fact does have the units of m/s if E and B are in units of volts per meter and teslas, respectively.

Determine the Concept Substituting the SI units for E and B yields:

$$\frac{\dfrac{N}{C}}{\dfrac{N}{A \cdot m}} = \frac{A \cdot m}{C} == \frac{\dfrac{C}{s} \cdot m}{C} = \boxed{\frac{m}{s}}$$

The Force Exerted by a Magnetic Field

23   ••   A 10-cm long straight wire is parallel with the *x* axis and carries a current of 2.0 A in the +*x* direction. The force on this wire due to the presence of a magnetic field $\vec{B}$ is $3.0\ \mathrm{N}\hat{j}+2.0\ \mathrm{N}\hat{k}$. If this wire is rotated so that it is parallel with the y axis with the current in the $+y$ direction, the force on the wire becomes $-3.0\ \mathrm{N}\hat{i}-2.0\ \mathrm{N}\hat{k}$. Determine the magnetic field $\vec{B}$.

Picture the Problem We can use the information given in the 1st and 2nd sentences to obtain an expression containing the components of the magnetic field $\vec{B}$. We can then use the information in the 1st and 3rd sentences to obtain a second equation in these components that we can solve simultaneously for the components of $\vec{B}$.

Express the magnetic field $\vec{B}$ in terms of its components:

$$\vec{B}=B_x\hat{i}+B_y\hat{j}+B_z\hat{k} \qquad (1)$$

Express $\vec{F}$ in terms of $\vec{B}$:

$$\vec{F}=I\vec{\ell}\times\vec{B}=(2.0\,\mathrm{A})\!\left[(0.10\,\mathrm{m})\hat{i}\right]\times\left(B_x\hat{i}+B_y\hat{j}+B_z\hat{k}\right)$$
$$=(0.20\,\mathrm{A}\cdot\mathrm{m})\hat{i}\times\left(B_x\hat{i}+B_y\hat{j}+B_z\hat{k}\right)=-(0.20\,\mathrm{A}\cdot\mathrm{m})B_z\hat{j}+(0.20\,\mathrm{A}\cdot\mathrm{m})B_y\hat{k}$$

Equate the components of this expression for $\vec{F}$ with those given in the second sentence of the statement of the problem to obtain:

$$-(0.20\,\mathrm{A}\cdot\mathrm{m})B_z=3.0\,\mathrm{N}$$
and
$$(0.20\,\mathrm{A}\cdot\mathrm{m})B_y=2.0\,\mathrm{N}$$

Noting that B_x is undetermined, solve for B_z and B_y:

$$B_z=-15\,\mathrm{T} \text{ and } B_y=10\,\mathrm{T}$$

When the wire is rotated so that the current flows in the positive y direction:

$$\vec{F}=I\vec{\ell}\times\vec{B}=(2.0\,\mathrm{A})\!\left[(0.10\,\mathrm{m})\hat{j}\right]\times\left(B_x\hat{i}+B_y\hat{j}+B_z\hat{k}\right)$$
$$=(0.20\,\mathrm{A}\cdot\mathrm{m})\hat{j}\times\left(B_x\hat{i}+B_y\hat{j}+B_z\hat{k}\right)=(0.20\,\mathrm{A}\cdot\mathrm{m})B_z\hat{i}-(0.20\,\mathrm{A}\cdot\mathrm{m})B_x\hat{k}$$

Equate the components of this expression for $\vec{F}$ with those given in the third sentence of the problem statement to obtain:

$$(0.20\,\mathrm{A}\cdot\mathrm{m})B_x=-2.0\,\mathrm{N}$$
and
$$-(0.20\,\mathrm{A}\cdot\mathrm{m})B_z=-3.0\,\mathrm{N}$$

Solve for B_x and B_z to obtain:

$B_x=10\,\mathrm{T}$ and, in agreement with our results above, $B_z=-15\,\mathrm{T}$

Substitute in equation (1) to obtain: $\vec{B} = \boxed{(10\,\text{T})\hat{i} + (10\,\text{T})\hat{j} - (15\,\text{T})\hat{k}}$

25 ••   A current-carrying wire is bent into a closed semicircular loop of radius R that lies in the xy plane (Figure 26-34). The wire is in a uniform magnetic field that is in the $+z$ direction, as shown. Verify that the force acting on the loop is zero.

Picture the Problem With the current in the direction indicated and the magnetic field in the z direction, pointing out of the plane of the page, the force is in the radial direction and we can integrate the element of force dF acting on an element of length $d\ell$ between $\theta = 0$ and π to find the force acting on the semicircular portion of the loop and use the expression for the force on a current-carrying wire in a uniform magnetic field to find the force on the straight segment of the loop.

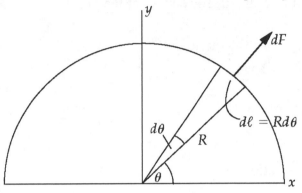

Express the net force acting on the semicircular loop of wire:	$\vec{F} = \vec{F}_{\substack{\text{semicircular}\\\text{loop}}} + \vec{F}_{\substack{\text{straight}\\\text{segment}}}$ (1)
Express the force acting on the straight segment of the loop:	$\vec{F}_{\substack{\text{straight}\\\text{segment}}} = I\vec{\ell} \times \vec{B} = 2RI\hat{i} \times B\hat{k} = -2RIB\hat{j}$
Express the force dF acting on the element of the wire of length $d\ell$:	$dF = Id\ell B = IRBd\theta$
Express the x and y components of dF:	$dF_x = dF\cos\theta$ and $dF_y = dF\sin\theta$
Because, by symmetry, the x component of the force is zero, we can integrate the y component to find the force on the wire:	$dF_y = IRB\sin\theta\,d\theta$ and $\vec{F}_{\substack{\text{semicircular}\\\text{loop}}} = F_y\hat{j} = \left(RIB\int_0^\pi \sin\theta\,d\theta\right)\hat{j}$ $= 2RIB\hat{j}$

Substitute in equation (1) to obtain: $\vec{F} = 2RIB\hat{j} - 2RIB\hat{j} = \boxed{0}$

Motion of a Point Charge in a Magnetic Field

27 •   A proton moves in a 65-cm-radius circular orbit that is perpendicular to a uniform magnetic field of magnitude 0.75 T. (*a*) What is the orbital period for the motion? (*b*) What is the speed of the proton? (*c*) What is the kinetic energy of the proton?

Picture the Problem We can apply Newton's 2nd law to the orbiting proton to relate its speed to its radius. We can then use $T = 2\pi r/v$ to find its period. In Part (*b*) we can use the relationship between T and v to determine v. In Part (*c*) we can use its definition to find the kinetic energy of the proton.

(*a*) Relate the period T of the motion of the proton to its orbital speed v:

$$T = \frac{2\pi r}{v} \qquad (1)$$

Apply Newton's 2nd law to the proton to obtain:

$$qvB = m\frac{v^2}{r} \Rightarrow r = \frac{mv}{qB}$$

Substitute for r in equation (1) and simplify to obtain:

$$T = \frac{2\pi m}{qB}$$

Substitute numerical values and evaluate T:

$$T = \frac{2\pi(1.673\times10^{-27}\text{ kg})}{(1.602\times10^{-19}\text{ C})(0.75\text{ T})} = 87.4\text{ ns}$$

$$= \boxed{87\text{ ns}}$$

(*b*) From equation (1) we have:

$$v = \frac{2\pi r}{T}$$

Substitute numerical values and evaluate v:

$$v = \frac{2\pi(0.65\text{ m})}{87.4\text{ ns}} = 4.67\times10^7\text{ m/s}$$

$$= \boxed{4.7\times10^7\text{ m/s}}$$

(*c*) Using its definition, express and evaluate the kinetic energy of the proton:

$$K = \tfrac{1}{2}mv^2 = \tfrac{1}{2}(1.673\times10^{-27}\text{ kg})(4.67\times10^7\text{ m/s})^2 = 1.82\times10^{-12}\text{ J}\times\frac{1\text{eV}}{1.602\times10^{-19}\text{ J}}$$

$$= \boxed{11\text{MeV}}$$

31 •• A beam of particles with velocity $\vec{v}$ enters a region that has a uniform magnetic field $\vec{B}$ in the $+x$ direction. Show that when the x component of the displacement of one of the particles is $2\pi(m/qB)v\cos\theta$, where θ is the angle between $\vec{v}$ and $\vec{B}$, the velocity of the particle is in the same direction as it was when the particle entered the field.

Picture the Problem The particle's velocity has a component v_1 parallel to $\vec{B}$ and a component v_2 normal to $\vec{B}$. $v_1 = v\cos\theta$ and is constant, whereas $v_2 = v\sin\theta$, being normal to $\vec{B}$, will result in a magnetic force acting on the beam of particles and circular motion perpendicular to $\vec{B}$. We can use the relationship between distance, rate, and time and Newton's 2nd law to express the distance the particle moves in the direction of the field during one period of the motion.

Express the distance moved in the direction of $\vec{B}$ by the particle during one period:

$$x = v_1 T \tag{1}$$

Express the period of the circular motion of the particles in the beam:

$$T = \frac{2\pi r}{v_2} \tag{2}$$

Apply Newton's 2nd law to a particle in the beam to obtain:

$$qv_2 B = m\frac{v_2^2}{r} \Rightarrow v_2 = \frac{qBr}{m}$$

Substituting for v_2 in equation (2) and simplifying yields:

$$T = \frac{2\pi r}{\dfrac{qBr}{m}} = \frac{2\pi m}{qB}$$

Because $v_1 = v\cos\theta$, equation (1) becomes:

$$x = (v\cos\theta)\left(\frac{2\pi m}{qB}\right) = \boxed{2\pi\left(\frac{m}{qB}\right)v\cos\theta}$$

33 •• Suppose that in Figure 26-35, the magnetic field has a magnitude of 60 mT, the distance d is 40 cm, and θ is 24°. Find the speed v at which a particle enters the region and the exit angle ϕ if the particle is a (*a*) proton and (*b*) deuteron. Assume that $m_d = 2m_p$.

Picture the Problem The trajectory of the proton is shown to the right. We know that, because the proton enters the uniform field perpendicularly to the field, its trajectory while in the field will be circular. We can use symmetry considerations to determine ϕ. The application of Newton's 2nd law to the proton and deuteron while they are in the uniform magnetic field will allow us to determine the values of v_p and v_d.

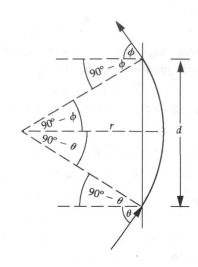

(a) From symmetry, it is evident that the angle θ in Figure 26-35 equals the angle ϕ:

$\phi = \boxed{24°}$

Apply $\sum F_{radial} = ma_c$ to the proton while it is in the magnetic field to obtain:

$$q_p v_p B = m_p \frac{v_p^2}{r_p} \Rightarrow v_p = \frac{q_p r_p B}{m_p} \quad (1)$$

Use trigonometry to obtain:

$$\sin(90° - \theta) = \sin 66° = \frac{d/2}{r}$$

Solving for r yields:

$$r = \frac{d}{2\sin 66°}$$

Substituting for r in equation (1) and simplifying yields:

$$v_p = \frac{q_p B d}{2 m_p \sin 66°} \quad (2)$$

Substitute numerical values and evaluate v_p:

$$v_p = \frac{\left(1.602 \times 10^{-19}\,\text{C}\right)\left(60\,\text{mT}\right)\left(0.40\,\text{m}\right)}{2\left(1.673 \times 10^{-27}\,\text{kg}\right)\sin 66°}$$

$$= \boxed{1.3 \times 10^6\,\text{m/s}}$$

(b) From symmetry, it is evident that the angle θ in Figure 26-35 equals the angle ϕ:

$\phi = \boxed{24°}$ independently of whether the particles are protons or deuterons.

For deuterons equation (2) becomes:

$$v_d = \frac{q_d B d}{2 m_d \sin 66°}$$

Because $m_d = 2m_p$ and $q_d = q_p$:

$$v_d \approx \frac{q_p Bd}{2(2m_p)\sin 66°} = \frac{q_p Bd}{4m_p \sin 66°}$$

Substitute numerical values and evaluate v_d:

$$v_d = \frac{(1.602 \times 10^{-19}\,\text{C})(60\,\text{mT})(0.40\,\text{m})}{4(1.673 \times 10^{-27}\,\text{kg})\sin 66°}$$

$$= \boxed{6.3 \times 10^5 \text{ m/s}}$$

Applications of the Magnetic Force Acting on Charged Particles

35 •   A velocity selector has a magnetic field that has a magnitude equal to 0.28 T and is perpendicular to an electric field that has a magnitude equal to 0.46 MV/m. (*a*) What must the speed of a particle be for that particle to pass through the velocity selector undeflected? What kinetic energy must (*b*) protons and (*c*) electrons have in order to pass through the velocity selector undeflected?

Picture the Problem Suppose that, for positively charged particles, their motion is from left to right through the velocity selector and the electric field is upward. Then the magnetic force must be downward and the magnetic field out of the page. We can apply the condition for translational equilibrium to relate v to E and B. In (*b*) and (*c*) we can use the definition of kinetic energy to find the energies of protons and electrons that pass through the velocity selector undeflected.

(*a*) Apply $\sum F_y = 0$ to the particle to obtain:

$$F_{elec} - F_{mag} = 0$$

or

$$qE - qvB = 0 \Rightarrow v = \frac{E}{B}$$

Substitute numerical values and evaluate v:

$$v = \frac{0.46\,\text{MV/m}}{0.28\,\text{T}} = 1.64 \times 10^6 \text{ m/s}$$

$$= \boxed{1.6 \times 10^6 \text{ m/s}}$$

(*b*) The kinetic energy of protons passing through the velocity selector undeflected is:

$$K_p = \tfrac{1}{2} m_p v^2$$

$$= \tfrac{1}{2}(1.673 \times 10^{-27}\,\text{kg})(1.64 \times 10^6 \text{ m/s})^2$$

$$= 2.26 \times 10^{-15}\,\text{J} \times \frac{1\,\text{eV}}{1.602 \times 10^{-19}\,\text{J}}$$

$$= \boxed{14\,\text{keV}}$$

(c) The kinetic energy of electrons passing through the velocity selector undeflected is:

$$K_e = \tfrac{1}{2} m_e v^2$$

$$= \tfrac{1}{2}\left(9.109 \times 10^{-31}\,_{\text{kg}}\right)\left(1.64 \times 10^6\,\text{m/s}\right)^2$$

$$= 1.23 \times 10^{-18}\,\text{J} \times \frac{1\text{eV}}{1.602 \times 10^{-19}\,\text{J}}$$

$$= \boxed{7.7\,\text{eV}}$$

39 ••   In a mass spectrometer, a singly ionized ^{24}Mg ion has a mass equal to 3.983×10^{-26} kg and is accelerated through a 2.50-kV potential difference. It then enters a region where it is deflected by a magnetic field of 557 G. (a) Find the radius of curvature of the ion's orbit. (b) What is the difference in the orbital radii of the ^{26}Mg and ^{24}Mg ions? Assume that their mass ratio is 26:24.

Picture the Problem We can apply Newton's 2nd law to an ion in the magnetic field to obtain an expression for r as a function of m, v, q, and B and use the work-kinetic energy theorem to express the kinetic energy in terms of the potential difference through which the ion has been accelerated. Eliminating v between these equations will allow us to express r in terms of m, q, B, and ΔV.

Apply Newton's 2nd law to an ion in the magnetic field of the mass spectrometer:

$$qvB = m\frac{v^2}{r} \Rightarrow r = \frac{mv}{qB} \qquad (1)$$

Apply the work-kinetic energy theorem to relate the speed of an ion as it enters the magnetic field to the potential difference through which it has been accelerated:

$$q\Delta V = \tfrac{1}{2}mv^2 \Rightarrow v = \sqrt{\frac{2q\Delta V}{m}}$$

Substitute for v in equation (1) and simplify to obtain:

$$r = \frac{m}{qB}\sqrt{\frac{2q\Delta V}{m}} = \sqrt{\frac{2m\Delta V}{qB^2}} \qquad (2)$$

(a) Substitute numerical values and evaluate equation (2) for ^{24}Mg :

$$r_{24} = \sqrt{\frac{2\left(3.983 \times 10^{-26}\,\text{kg}\right)\left(2.50\,\text{kV}\right)}{\left(1.602 \times 10^{-19}\,\text{C}\right)\left(557 \times 10^{-4}\,\text{T}\right)^2}}$$

$$= \boxed{63.3\,\text{cm}}$$

(b) Express the difference in the radii for ^{24}Mg and ^{26}Mg:

$$\Delta r = r_{26} - r_{24}$$

Substituting for r_{26} and r_{24} and simplifying yields:

$$\Delta r = \sqrt{\frac{2m_{26}\Delta V}{qB^2}} - \sqrt{\frac{2m_{24}\Delta V}{qB^2}} = \frac{1}{B}\sqrt{\frac{2\Delta V}{q}}\left(\sqrt{m_{26}} - \sqrt{m_{24}}\right)$$

$$= \frac{1}{B}\sqrt{\frac{2\Delta V}{q}}\left(\sqrt{\frac{26}{24}m_{24}} - \sqrt{m_{24}}\right) = \frac{1}{B}\sqrt{\frac{2\Delta V m_{24}}{q}}\left(\sqrt{\frac{26}{24}} - 1\right)$$

Substitute numerical values and evaluate Δr:

$$\Delta r = \frac{1}{557\times10^{-4}\,\text{T}}\sqrt{\frac{2(2.50\,\text{kV})(3.983\times10^{-26}\,\text{kg})}{1.602\times10^{-19}\,\text{C}}}\left(\sqrt{\frac{26}{24}} - 1\right) = \boxed{2.58\,\text{cm}}$$

43 •• A cyclotron for accelerating protons has a magnetic field strength of 1.4 T and a radius of 0.70 m. (a) What is the cyclotron's frequency? (b) Find the kinetic energy of the protons when they emerge. (c) How will your answers change if deuterons are used instead of protons?

Picture the Problem We can express the cyclotron frequency in terms of the maximum orbital radius and speed of the protons/deuterons. By applying Newton's 2nd law, we can relate the radius of the particle's orbit to its speed and, hence, express the cyclotron frequency as a function of the particle's mass and charge and the cyclotron's magnetic field. In Part (b) we can use the definition of kinetic energy and their maximum speed to find the maximum energy of the emerging protons.

(a) Express the cyclotron frequency in terms of the proton's orbital speed and radius:

$$f = \frac{1}{T} = \frac{1}{2\pi r/v} = \frac{v}{2\pi r} \qquad (1)$$

Apply Newton's 2nd law to a proton in the magnetic field of the cyclotron:

$$qvB = m\frac{v^2}{r} \Rightarrow r = \frac{mv}{qB} \qquad (2)$$

Substitute for r in equation (1) and simplify to obtain:

$$f = \frac{qBv}{2\pi mv} = \frac{qB}{2\pi m} \qquad (3)$$

| Substitute numerical values and evaluate f: | $f = \dfrac{(1.602 \times 10^{-19}\,\text{C})(1.4\,\text{T})}{2\pi(1.673 \times 10^{-27}\,\text{kg})} = 21.3\,\text{MHz}$ |
| | $= \boxed{21\,\text{MHz}}$ |

| (b) Express the maximum kinetic energy of a proton: | $K_{max} = \tfrac{1}{2} m v_{max}^2$ |

| From equation (2), v_{max} is given by: | $v_{max} = \dfrac{qBr_{max}}{m}$ |

| Substitute for v_{max} and simplify to obtain: | $K_{max} = \tfrac{1}{2} m \left(\dfrac{qBr_{max}}{m} \right)^2 = \tfrac{1}{2} \left(\dfrac{q^2 B^2}{m} \right) r_{max}^2$ |

Substitute numerical values and evaluate K_{max}:

$$K_{max} = \tfrac{1}{2} \left(\dfrac{(1.602 \times 10^{-19}\,\text{C})^2 (1.4\,\text{T})^2}{1.673 \times 10^{-27}\,\text{kg}} \right) (0.7\,\text{m})^2 = 7.37 \times 10^{-12}\,\text{J} \times \dfrac{1\,\text{eV}}{1.602 \times 10^{-19}\,\text{J}}$$

$$= 46.0\,\text{MeV} = \boxed{46\,\text{MeV}}$$

| (c) From equation (3) we see that doubling m halves f: | $f_{deuterons} = \tfrac{1}{2} f_{protons} = \boxed{11\,\text{MHz}}$ |

| From our expression for K_{max} we see that doubling m halves K: | $K_{deuterons} = \tfrac{1}{2} K_{protons} = \boxed{23\,\text{MeV}}$ |

Torques on Current Loops, Magnets, and Magnetic Moments

47 •   A small circular coil consisting of 20 turns of wire lies in a region with a uniform magnetic field whose magnitude is 0.50 T. The arrangement is such that the normal to the plane of the coil makes an angle of 60° with the direction of the magnetic field. The radius of the coil is 4.0 cm, and the wire carries a current of 3.0 A. (a) What is the magnitude of the magnetic moment of the coil? (b) What is the magnitude of the torque exerted on the coil?

Picture the Problem We can use the definition of the magnetic moment of a coil to evaluate μ and the expression for the torque exerted on the coil $\vec{\tau} = \vec{\mu} \times \vec{B}$ to find the magnitude of τ.

(a) Using its definition, express the magnetic moment of the coil:

$$\mu = NIA = NI\pi r^2$$

Substitute numerical values and evaluate μ:

$$\mu = (20)(3.0\,\text{A})\pi(0.040\,\text{m})^2$$
$$= 0.302\,\text{A}\cdot\text{m}^2 = \boxed{0.30\,\text{A}\cdot\text{m}^2}$$

(b) Express the magnitude of the torque exerted on the coil:

$$\tau = \mu B \sin\theta$$

Substitute numerical values and evaluate τ:

$$\tau = (0.302\,\text{A}\cdot\text{m}^2)(0.50\,\text{T})\sin 60°$$
$$= \boxed{0.13\,\text{N}\cdot\text{m}}$$

49 •   A current-carrying wire is in the shape of a square of edge-length 6.0 cm. The square lies in the $z = 0$ plane. The wire carries a current of 2.5 A. What is the magnitude of the torque on the wire if it is in a region with a uniform magnetic field of magnitude 0.30 T that points in the (a) +z direction and (b) +x direction?

Picture the Problem We can use $\vec{\tau} = \vec{\mu} \times \vec{B}$ to find the torque on the coil in the two orientations of the magnetic field.

Express the torque acting on the coil:

$$\vec{\tau} = \vec{\mu} \times \vec{B}$$

Express the magnetic moment of the coil:

$$\vec{\mu} = \pm IA\hat{k} = \pm IL^2\hat{k}$$

(a) Evaluate $\vec{\tau}$ for $\vec{B}$ in the +z direction:

$$\vec{\tau} = \pm IL^2\hat{k} \times B\hat{k} = \pm IL^2 B(\hat{k}\times\hat{k}) = \boxed{0}$$

(b) Evaluate $\vec{\tau}$ for $\vec{B}$ in the +x direction:

$$\vec{\tau} = \pm IL^2\hat{k} \times B\hat{i} = \pm IL^2 B(\hat{k}\times\hat{i})$$
$$= \pm(2.5\,\text{A})(0.060\,\text{m})^2(0.30\,\text{T})\hat{j}$$
$$= \pm(2.7\,\text{mN}\cdot\text{m})\hat{j}$$
and
$$|\vec{\tau}| = \boxed{2.7\times10^{-3}\,\text{N}\cdot\text{m}}$$

53 ••   For the coil in Problem 52 the magnetic field is now
$\vec{B} = 2.0 \text{ T}\hat{j}$. Find the torque exerted on the coil when $\hat{n}$ is equal to (*a*) $\hat{i}$, (*b*) $\hat{j}$,
(*c*) $-\hat{j}$, and (*d*) $\dfrac{\hat{i}}{\sqrt{2}} + \dfrac{\hat{j}}{\sqrt{2}}$.

Picture the Problem We can use the right-hand rule for determining the direction
of $\hat{n}$ to establish the orientation of the coil for value of $\hat{n}$ and $\vec{\tau} = \vec{\mu} \times \vec{B}$ to find the
torque exerted on the coil in each orientation.

(*a*) The orientation of the coil
is shown to the right:

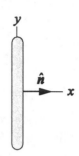

Evaluate $\vec{\tau}$ for $\vec{B} = 2.0 \text{ T}\,\hat{j}$ and
$\hat{n} = \hat{i}$:

$\vec{\tau} = \vec{\mu} \times \vec{B} = NIA\hat{n} \times \vec{B}$
$= (50)(1.75 \text{ A})(48.0 \text{ cm}^2)\hat{i} \times (2.0 \text{ T})\hat{j}$
$= (0.840 \text{ N} \cdot \text{m})(\hat{i} \times \hat{j}) = (0.840 \text{ N} \cdot \text{m})\hat{k}$
$= \boxed{(0.84 \text{ N} \cdot \text{m})\hat{k}}$

(*b*) The orientation of the coil is
shown to the right:

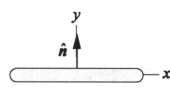

Evaluate $\vec{\tau}$ for $\vec{B} = 2.0 \text{ T}\,\hat{j}$ and
$\hat{n} = \hat{j}$:

$\vec{\tau} = \vec{\mu} \times \vec{B} = NIA\hat{n} \times \vec{B}$
$= (50)(1.75 \text{ A})(48.0 \text{ cm}^2)\hat{j} \times (2.0 \text{ T})\hat{j}$
$= (0.840 \text{ N} \cdot \text{m})(\hat{j} \times \hat{j})$
$= \boxed{0}$

(*c*) The orientation of the coil is
shown to the right:

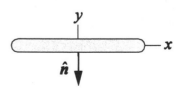

Evaluate $\vec{\tau}$ for $\vec{B} = 2.0\,\text{T}\,\hat{j}$ and $\hat{n} = -\hat{j}$:

$$\vec{\tau} = \vec{\mu} \times \vec{B} = NIA\hat{n} \times \vec{B}$$
$$= -(50)(1.75\,\text{A})(48.0\,\text{cm}^2)\hat{j} \times (2.0\,\text{T})\hat{j}$$
$$= (-0.840\,\text{N} \cdot \text{m})(\hat{j} \times \hat{j})$$
$$= \boxed{0}$$

(d) The orientation of the coil is shown to the right:

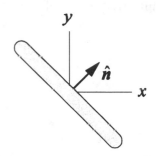

Evaluate $\vec{\tau}$ for $\vec{B} = 2.0\,\text{T}\,\hat{j}$ and $\hat{n} = (\hat{i} + \hat{j})/\sqrt{2}$:

$$\vec{\tau} = \vec{\mu} \times \vec{B} = NIA\hat{n} \times \vec{B}$$
$$= \frac{(50)(1.75\,\text{A})(48.0\,\text{cm}^2)}{\sqrt{2}}(\hat{i} + \hat{j}) \times (2.0\,\text{T})\hat{j}$$
$$= (0.594\,\text{N} \cdot \text{m})(\hat{i} \times \hat{j})$$
$$\quad + (0.594\,\text{N} \cdot \text{m})(\hat{j} \times \hat{j})$$
$$= \boxed{(0.59\,\text{N} \cdot \text{m})\hat{k}}$$

57 •• A particle that has a charge q and a mass m moves with angular velocity ω in a circular path of radius r. (a) Show that the average current created by this moving particle is $\omega q/(2\pi)$ and that the magnetic moment of its orbit has a magnitude of $\frac{1}{2}q\omega r^2$. (b) Show that the angular momentum of this particle has the magnitude of $mr^2\omega$ and that the magnetic moment and angular momentum vectors are related by $\vec{\mu} = (q/2m)\vec{L}$, where $\vec{L}$ is the angular momentum about the center of the circle.

Picture the Problem We can use the definition of current and the relationship between the frequency of the motion and its period to show that $I = q\omega/2\pi$. We can use the definition of angular momentum and the moment of inertia of a point particle to show that the magnetic moment has the magnitude $\mu = \frac{1}{2}q\omega r^2$. Finally, we can express the ratio of μ to L and the fact that $\vec{\mu}$ and $\vec{L}$ are both parallel to $\vec{\omega}$ to conclude that $\vec{\mu} = (q/2m)\vec{L}$.

(a) Using its definition, relate the average current to the charge passing a point on the circumference of the circle in a given period of time:

$$I = \frac{\Delta q}{\Delta t} = \frac{q}{T} = qf$$

Relate the frequency of the motion to the angular frequency of the particle:

$$f = \frac{\omega}{2\pi}$$

Substitute for f to obtain:

$$I = \boxed{\frac{q\omega}{2\pi}}$$

From the definition of the magnetic moment we have:

$$\mu = IA = \left(\frac{q\omega}{2\pi}\right)\left(\pi r^2\right) = \boxed{\tfrac{1}{2}q\omega r^2}$$

(b) Express the angular momentum of the particle:

$$L = I\omega$$

The moment of inertia of the particle is:

$$I = mr^2$$

Substituting for I yields:

$$L = \left(mr^2\right)\omega = \boxed{mr^2\omega}$$

Express the ratio of μ to L and simplify to obtain:

$$\frac{\mu}{L} = \frac{\tfrac{1}{2}q\omega r^2}{mr^2\omega} = \frac{q}{2m} \Rightarrow \mu = \frac{q}{2m}L$$

Because $\vec{\mu}$ and $\vec{L}$ are both parallel to $\vec{\omega}$:

$$\vec{\mu} = \boxed{\frac{q}{2m}\vec{L}}$$

59 •••   A uniform non-conducting thin rod of mass m and length L has a uniform charge per unit length λ and rotates with angular speed ω about an axis through one end and perpendicular to the rod. (a) Consider a small segment of the rod of length dx and charge $dq = \lambda dr$ at a distance r from the pivot (Figure 26-40). Show that the average current created by this moving segment is $\omega dq/(2\pi)$ and show that the magnetic moment of this segment is $\tfrac{1}{2}\lambda\omega r^2 dx$. (b) Use this to show that the magnitude of the magnetic moment of the rod is $\tfrac{1}{6}\lambda\omega L^3$. (c) Show that the magnetic moment $\vec{\mu}$ and angular momentum $\vec{L}$ are related by $\vec{\mu} = Q/2M\vec{L}$, where Q is the total charge on the rod.

Picture the Problem We can follow the step-by-step outline provided in the problem statement to establish the given results.

(*a*) Express the magnetic moment of the rotating element of charge:

$$d\mu = A\,dI \qquad\qquad (1)$$

The area enclosed by the rotating element of charge is:

$$A = \pi x^2$$

Express dI in terms of dq and Δt:

$$dI = \frac{dq}{\Delta t} = \frac{\lambda\,dx}{\Delta t} \text{ where } \Delta t \text{ is the time}$$

required for one revolution.

The time Δt required for one revolution is:

$$\Delta t = \frac{1}{f} = \frac{2\pi}{\omega}$$

Substitute for Δt and simplify to obtain:

$$dI = \frac{\lambda\omega}{2\pi}\,dx$$

Substituting for dI in equation (1) and simplifying yields:

$$d\mu = \left(\pi x^2\right)\left(\frac{\lambda\omega}{2\pi}\,dx\right) = \boxed{\tfrac{1}{2}\lambda\omega x^2\,dx}$$

(*b*) Integrate $d\mu$ from $x = 0$ to $x = L$ to obtain:

$$\mu = \tfrac{1}{2}\lambda\omega\int_0^L x^2\,dx = \boxed{\tfrac{1}{6}\lambda\omega L^3}$$

(*c*) Express the angular momentum of the rod:

$$L = I\omega$$

where L is the angular momentum of the rod and I is the moment of inertia of the rod with respect to the point about which it is rotating.

Express the moment of inertia of the rod with respect to an axis through its end:

$$I = \tfrac{1}{3}mL^2$$

where L is now the length of the rod.

Substitute to obtain:

$$L = \tfrac{1}{3}mL^2\omega$$

Divide the expression for μ by L to obtain:

$$\frac{\mu}{L} = \frac{\frac{1}{6}\lambda\omega L^3}{\frac{1}{3}mL^2\omega} = \frac{\lambda L}{2m}$$

or, because $Q = \lambda L$,

$$\mu = \frac{Q}{2m}L$$

Because $\vec{\omega}$ and $\vec{L} = I\vec{\omega}$ are parallel:

$$\boxed{\vec{\mu} = \frac{Q}{2M}\vec{L}}$$

61 •••   A spherical shell of radius R carries a constant surface charge density σ. The shell rotates about its diameter with angular speed ω. Find the magnitude of the magnetic moment of the rotating shell.

Picture the Problem We can use the result of Problem 57 to express μ as a function of Q, M, and L. We can then use the definitions of surface charge density and angular momentum to substitute for Q and L to obtain the magnetic moment of the rotating shell.

Express the magnetic moment of the spherical shell in terms of its mass, charge, and angular momentum:

$$\mu = \frac{Q}{2M}L$$

Use the definition of surface charge density to express the charge on the spherical shell:

$$Q = \sigma A = 4\pi\sigma R^2$$

Express the angular momentum of the spherical shell:

$$L = I\omega = \tfrac{2}{3}MR^2\omega$$

Substitute for L and simplify to obtain:

$$\mu = \left(\frac{4\pi\sigma R^2}{2M}\right)\left(\frac{2}{3}MR^2\omega\right) = \boxed{\tfrac{4}{3}\pi\sigma R^4\omega}$$

65 ••   The number density of free electrons in copper is 8.47×10^{22} electrons per cubic centimeter. If the metal strip in Figure 26-41 is copper and the current is 10.0 A, find (a) the drift speed v_d and (b) the potential difference $V_a - V_b$. Assume that the magnetic field strength is 2.00 T.

Picture the Problem We can use $I = nqv_dA$ to find the drift speed and $V_H = v_dBw$ to find the potential difference $V_a - V_b$.

(*a*) Express the current in the metal strip in terms of the drift speed of the electrons:

$$I = nqv_d A \Rightarrow v_d = \frac{I}{nqA}$$

Substitute numerical values and evaluate v_d:

$$v_d = \frac{10.0\,\text{A}}{\left(8.47\times10^{22}\,\text{cm}^{-3}\right)\left(1.602\times10^{-19}\,\text{C}\right)\left(2.00\,\text{cm}\right)\left(0.100\,\text{cm}\right)} = 3.685\times10^{-5}\,\text{m/s}$$

$$= \boxed{3.68\times10^{-5}\,\text{m/s}}$$

(*b*) The potential difference $V_a - V_b$ is the Hall voltage and is given by:

$$V_a - V_b = V_H = v_d B w$$

Substitute numerical values and evaluate $V_a - V_b$:

$$V_a - V_b = \left(3.685\times10^{-5}\,\text{m/s}\right)\left(2.00\,\text{T}\right)\left(2.00\,\text{cm}\right) = \boxed{1.47\,\mu\text{V}}$$

69 ••   Aluminum has a density of 2.7×10^3 kg/m^3 and a molar mass of 27 g/mol. The Hall coefficient of aluminum is $R = -0.30 \times 10^{-10}$ m^3/C. (See Problem 68 for the definition of R.) What is the number of conduction electrons per aluminum atom?

Picture the Problem We can determine the number of conduction electrons per atom from the quotient of the number density of charge carriers and the number of charge carriers per unit volume. Let the width of a slab of aluminum be w and its thickness t. We can use the definition of the Hall electric field in the slab, the expression for the Hall voltage across it, and the definition of current density to find n in terms of R and q and $n_a = \rho N_A / M$, to express n_a.

Express the number of electrons per atom N:

$$N = \frac{n}{n_a} \tag{1}$$

where n is the number density of charge carriers and n_a is the number of atoms per unit volume.

From the definition of the Hall coefficient we have:

$$R = \frac{E_y}{J_x B_z}$$

Express the Hall electric field in the slab:

$$E_y = \frac{V_H}{w}$$

The current density in the slab is:

$$J_x = \frac{I}{wt} = nqv_d$$

Substitute for E_y and J_x in the expression for R to obtain:

$$R = \frac{\frac{V_H}{w}}{nqv_d B_z} = \frac{V_H}{nqv_d w B_z}$$

Express the Hall voltage in terms of v_d, B, and w:

$$V_H = v_d B_z w$$

Substitute for V_H and simplify to obtain:

$$R = \frac{v_d B_z w}{nqv_d w B_z} = \frac{1}{nq} \Rightarrow n = \frac{1}{Rq} \quad (2)$$

Express the number of atoms n_a per unit volume:

$$n_a = \rho \frac{N_A}{M} \quad (3)$$

Substitute equations (2) and (3) in equation (1) to obtain:

$$N = \frac{M}{qR\rho N_A}$$

Substitute numerical values and evaluate N:

$$N = \frac{27\frac{\text{g}}{\text{mol}}}{\left(-1.602 \times 10^{-19}\,\text{C}\right)\left(-0.30 \times 10^{-10}\,\frac{\text{m}^3}{\text{C}}\right)\left(2.7 \times 10^3\,\frac{\text{kg}}{\text{m}^3}\right)\left(6.022 \times 10^{23}\,\frac{\text{atoms}}{\text{mol}}\right)}$$

$$\approx \boxed{4}$$

General Problems

73 ••   A particle of mass m and charge q enters a region where there is a uniform magnetic field $\vec{B}$ parallel with the x axis. The initial velocity of the particle is $\vec{v} = v_{0x}\hat{i} + v_{0y}\hat{j}$, so the particle moves in a helix. (a) Show that the radius of the helix is $r = mv_{0y}/qB$. (b) Show that the particle takes a time $\Delta t = 2\pi m/qB$ to complete each turn of the helix. (c) What is the x component of the displacement of the particle during time given in Part (b)?

Picture the Problem We can use $\vec{F} = q\vec{v} \times \vec{B}$ to show that motion of the particle in the x direction is not affected by the magnetic field. The application of Newton's 2nd law to motion of the particle in yz plane will lead us to the result that $r = mv_{0y}/qB$. By expressing the period of the motion in terms of v_{0y} we can show that the time for one complete orbit around the helix is $t = 2\pi m/qB$.

(a) Express the magnetic force acting on the particle:

$$\vec{F} = q\vec{v} \times \vec{B}$$

Substitute for $\vec{v}$ and $\vec{B}$ and simplify to obtain:

$$\vec{F} = q\left(v_{0x}\hat{i} + v_{0y}\hat{j}\right) \times B\hat{i}$$
$$= qv_{0x}B\left(\hat{i} \times \hat{i}\right) + qv_{0y}B\left(\hat{j} \times \hat{i}\right)$$
$$= 0 - qv_{0y}B\hat{k} = -qv_{0y}B\hat{k}$$

i.e., the motion in the direction of the magnetic field (the x direction) is not affected by the field.

Apply Newton's 2nd law to the particle in the plane perpendicular to $\hat{i}$ (i.e., the yz plane):

$$qv_{0y}B = m\frac{v_{0y}^2}{r} \qquad (1)$$

Solving for r yields:

$$\boxed{r = \frac{mv_{0y}}{qB}}$$

(b) Relate the time for one orbit around the helix to the particle's orbital speed:

$$\Delta t = \frac{2\pi r}{v_{0y}}$$

Solve equation (1) for v_{0y}:

$$v_{0y} = \frac{qBr}{m}$$

Substitute for v_{0y} and simplify to obtain:

$$\Delta t = \frac{2\pi r}{\frac{qBr}{m}} = \boxed{\frac{2\pi m}{qB}}$$

(c) Because, as was shown in Part (a), the motion in the direction of the magnetic field (the x direction) is not affected by the field, the x component of the displacement of the particle as a function of t is:

$$x(t) = v_{0x}t$$

For $t = \Delta t$:

$$x(\Delta t) = v_{0x}\left(\frac{2\pi m}{qB}\right) = \boxed{\frac{2\pi mv_{0x}}{qB}}$$

75 ••   Assume that the rails Problem 74 are frictionless but tilted upward so that they make an angle θ with the horizontal, and with the current source attached to the low end of the rails. The magnetic field is still directed vertically downward. (*a*) What minimum value of *B* is needed to keep the bar from sliding down the rails? (*b*) What is the acceleration of the bar if *B* is twice the value found in Part (*a*)?

Picture the Problem Note that with the rails tilted, $\vec{F}$ still points horizontally to the right (*I*, and hence $\vec{\ell}$, is out of the page). Choose a coordinate system in which down the incline is the positive *x* direction. Then we can apply a condition for translational equilibrium to find the vertical magnetic field $\vec{B}$ needed to keep the bar from sliding down the rails. In Part (*b*) we can apply Newton's 2nd law to find the acceleration of the crossbar when *B* is twice its value found in (*a*).

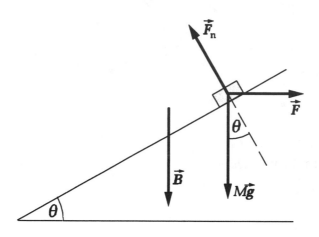

(*a*) Apply $\sum F_x = 0$ to the crossbar to obtain:

$$mg \sin \theta - I\ell B \cos \theta = 0$$

Solving for *B* yields:

$$B = \frac{mg}{I\ell} \tan \theta \text{ and } \vec{B} = \boxed{-\frac{mg}{I\ell} \tan \theta \, \hat{u}_v}$$

where $\hat{u}_v$ is a unit vector in the vertical direction.

(*b*) Apply Newton's 2nd law to the crossbar to obtain:

$$I\ell B' \cos \theta - mg \sin \theta = ma$$

Solving for *a* yields:

$$a = \frac{I\ell B'}{m} \cos \theta - g \sin \theta$$

Substitute $B' = 2B$ and simplify to obtain:

$$a = \frac{2I\ell \frac{mg}{I\ell} \tan\theta}{m} \cos\theta - g\sin\theta$$

$$= 2g\sin\theta - g\sin\theta = \boxed{g\sin\theta}$$

Note that the direction of the acceleration is up the incline.

Chapter 27
Sources of the Magnetic Field

Conceptual Problems

5 •   Discuss the differences and similarities between Gauss's law for magnetism and Gauss's law for electricity.

Determine the Concept Both tell you about the respective fluxes through closed surfaces. In the electrical case, the flux is proportional to the net charge enclosed. In the magnetic case, the flux is always zero because there is no such thing as magnetic charge (a magnetic monopole). The source of the magnetic field is NOT the equivalent of electric charge; that is, it is NOT a thing called magnetic charge, but rather it is moving electric charges.

7 •   You are facing directly into one end of a long solenoid and the magnetic field inside of the solenoid points away from you. From your perspective, is the direction of the current in the solenoid coils clockwise or counterclockwise? Explain your answer.

Determine the Concept Application of the right-hand rule leads one to conclude that the current is clockwise.

11 •   Of the four gases listed in Table 27-1, which are diamagnetic and which are paramagnetic?

Determine the Concept H_2, CO_2, and N_2 are diamagnetic ($\chi_m < 0$); O_2 is paramagnetic ($\chi_m > 0$).

The Magnetic Field of Moving Point Charges

13 •   At time $t = 0$, a particle has a charge of 12 μC, is located in the $z = 0$ plane at $x = 0$, $y = 2.0$ m, and has a velocity equal to 30 m/s $\hat{i}$. Find the magnetic field in the $z = 0$ plane at (*a*) the origin, (*b*) $x = 0$, $y = 1.0$ m, (*c*) $x = 0$, $y = 3.0$ m, and (*d*) $x = 0$, $y = 4.0$ m.

Picture the Problem We can substitute for $\vec{v}$ and q in the equation describing the magnetic field of the moving charged particle ($\vec{B} = \dfrac{\mu_0}{4\pi} \dfrac{q\vec{v} \times \hat{r}}{r^2}$), evaluate r and $\hat{r}$ for each of the given points of interest, and then find $\vec{B}$.

The magnetic field of the moving
charged particle is given by:

$$\vec{B} = \frac{\mu_0}{4\pi} \frac{q\vec{v} \times \hat{r}}{r^2}$$

Substitute numerical values and simplify to obtain:

$$\vec{B} = \left(10^{-7}\,\text{N/A}^2\right)\left(12\,\mu\text{C}\right)\frac{\left(30\,\text{m/s}\right)\hat{i}\times\hat{r}}{r^2}$$

$$= \left(36.0\,\text{pT}\cdot\text{m}^2\right)\frac{\hat{i}\times\hat{r}}{r^2}$$

(*a*) Find r and $\hat{r}$ for the particle at (0, 2.0 m) and the point of interest at the origin:

$$\vec{r} = -(2.0\,\text{m})\hat{j}\,,\; r = 2.0\,\text{m}\,,\; \text{and}\; \hat{r} = -\hat{j}$$

Evaluating $\vec{B}(0,0)$ yields:

$$\vec{B}(0,0) = \left(36.0\,\text{pT}\cdot\text{m}^2\right)\frac{\hat{i}\times\left(-\hat{j}\right)}{(2.0\,\text{m})^2}$$

$$= \boxed{-(9.0\,\text{pT})\hat{k}}$$

(*b*) Find r and $\hat{r}$ for the particle at (0, 2.0 m) and the point of interest at (0, 1.0 m):

$$\vec{r} = -(1.0\,\text{m})\hat{j}\,,\; r = 1.0\,\text{m}\,,\; \text{and}\; \hat{r} = -\hat{j}$$

Evaluate $\vec{B}(0,1.0\,\text{m})$ to obtain:

$$\vec{B}(0,1.0\,\text{m}) = \left(36.0\,\text{pT}\cdot\text{m}^2\right)\frac{\hat{i}\times\left(-\hat{j}\right)}{(1.0\,\text{m})^2}$$

$$= \boxed{-(36\,\text{pT})\hat{k}}$$

(*c*) Find r and $\hat{r}$ for the particle at (0, 2.0 m) and the point of interest at (0, 3.0 m):

$$\vec{r} = (1.0\,\text{m})\hat{j}\,,\; r = 1.0\,\text{m}\,,\; \text{and}\; \hat{r} = \hat{j}$$

Evaluating $\vec{B}(0,3.0\,\text{m})$ yields:

$$\vec{B}(0,3.0\,\text{m}) = \left(36.0\,\text{pT}\cdot\text{m}^2\right)\frac{\hat{i}\times\hat{j}}{(1.0\,\text{m})^2}$$

$$= \boxed{(36\,\text{pT})\hat{k}}$$

(*d*) Find r and $\hat{r}$ for the particle at (0, 2.0 m) and the point of interest at (0, 4.0 m):

$$\vec{r} = (2.0\,\text{m})\hat{j}\,,\; r = 2.0\,\text{m}\,,\; \text{and}\; \hat{r} = \hat{j}$$

Evaluate $\vec{B}(0,4.0\,\text{m})$ to obtain:

$$\vec{B}(0,4.0\,\text{m}) = \left(36.0\,\text{pT}\cdot\text{m}^2\right)\frac{\hat{i}\times\hat{j}}{(2.0\,\text{m})^2}$$

$$= \boxed{(9.0\,\text{pT})\hat{k}}$$

The Magnetic Field Using the Biot–Savart Law

19 • A small current element at the origin has a length of 2.0 mm and carries a current of 2.0 A in the +z direction. Find the magnitude and direction of the magnetic field due to the current element at the point (0, 3.0 m, 4.0 m).

Picture the Problem We can substitute for I and $d\vec{\ell} \approx \Delta\vec{\ell}$ in the Biot-Savart law ($d\vec{B} = \dfrac{\mu_0}{4\pi}\dfrac{Id\vec{\ell}\times\hat{r}}{r^2}$), evaluate r and $\hat{r}$ for (0, 3.0 m, 4.0 m), and substitute to find $d\vec{B}$.

The Biot-Savart law for the given current element is given by:	$d\vec{B} = \dfrac{\mu_0}{4\pi}\dfrac{Id\vec{\ell}\times\hat{r}}{r^2}$

Substituting numerical values yields:

$$d\vec{B} = \left(1.0\times10^{-7}\ \text{N/A}^2\right)\frac{(2.0\,\text{A})(2.0\,\text{mm})\hat{k}\times\hat{r}}{r^2} = \left(0.400\,\text{nT}\cdot\text{m}^2\right)\frac{\hat{k}\times\hat{r}}{r^2}$$

Find r and $\hat{r}$ for the point whose coordinates are (0, 3.0 m, 4.0 m):	$\vec{r} = (3.0\,\text{m})\hat{j} + (4.0\,\text{m})\hat{k}$, $r = 5.0\,\text{m}$, and $\hat{r} = \dfrac{3}{5}\hat{j} + \dfrac{4}{5}\hat{k}$

Evaluate $d\vec{B}$ at (0, 3.0 m, 4.0 m):

$$d\vec{B}(0, 3.0\,\text{m}, 4.0\,\text{m}) = \left(0.400\,\text{nT}\cdot\text{m}^2\right)\frac{\hat{k}\times\left(\dfrac{3}{5}\hat{j} + \dfrac{4}{5}\hat{k}\right)}{(5.0\,\text{m})^2} = \boxed{-(9.6\,\text{pT})\hat{i}}$$

The Magnetic Field Due to Straight-Line Currents

25 •• If the currents are both in the −x direction, find the magnetic field at the following points on the y axis: (a) y = −3.0 cm, (b) y = 0, (c) y = +3.0 cm, and (d) y = +9.0 cm.

Picture the Problem Let + denote the wire (and current) at y = +6.0 cm and − the wire (and current) at y = −6.0 cm. We can use $B = \dfrac{\mu_0}{4\pi}\dfrac{2I}{R}$ to find the magnetic field due to each of the current-carrying wires and superimpose the magnetic fields due to the currents in these wires to find B at the given points on the y axis. We can apply the right-hand rule to find the direction of each of the fields and, hence, of $\vec{B}$.

(*a*) Express the resultant magnetic field at $y = -3.0$ cm:

$$\vec{B}(-3.0\,\text{cm}) = \vec{B}_+(-3.0\,\text{cm}) + \vec{B}_-(-3.0\,\text{cm}) \quad (1)$$

Find the magnitudes of the magnetic fields at $y = -3.0$ cm due to each wire:

$$B_+(-3.0\,\text{cm}) = (10^{-7}\,\text{T}\cdot\text{m/A})\frac{2(20\,\text{A})}{0.090\,\text{m}}$$

$$= 44.4\,\mu\text{T}$$

and

$$B_-(-3.0\,\text{cm}) = (10^{-7}\,\text{T}\cdot\text{m/A})\frac{2(20\,\text{A})}{0.030\,\text{m}}$$

$$= 133\,\mu\text{T}$$

Apply the right-hand rule to find the directions of $\vec{B}_+$ and $\vec{B}_-$:

$$\vec{B}_+(-3.0\,\text{cm}) = (44.4\,\mu\text{T})\hat{k}$$

and

$$\vec{B}_-(-3.0\,\text{cm}) = -(133\,\mu\text{T})\hat{k}$$

Substituting in equation (1) yields:

$$\vec{B}(-3.0\,\text{cm}) = (44.4\,\mu\text{T})\hat{k} - (133\,\mu\text{T})\hat{k}$$

$$= \boxed{-(89\,\mu\text{T})\hat{k}}$$

(*b*) Express the resultant magnetic field at $y = 0$:

$$\vec{B}(0) = \vec{B}_+(0) + \vec{B}_-(0)$$

Because $\vec{B}_+(0) = -\vec{B}_-(0)$:

$$\vec{B}(0) = \boxed{0}$$

(*c*) Proceed as in (*a*) to obtain:

$$\vec{B}_+(3.0\,\text{cm}) = (133\,\mu\text{T})\hat{k},$$
$$\vec{B}_-(3.0\,\text{cm}) = -(44.4\,\mu\text{T})\hat{k},$$
and
$$\vec{B}(3.0\,\text{cm}) = (133\,\mu\text{T})\hat{k} - (44.4\,\mu\text{T})\hat{k}$$

$$= \boxed{(89\,\mu\text{T})\hat{k}}$$

(*d*) Proceed as in (*a*) with $y = 9.0$ cm to obtain:

$$\vec{B}_+(9.0\,\text{cm}) = -(133\,\mu\text{T})\hat{k},$$
$$\vec{B}_-(9.0\,\text{cm}) = -(26.7\,\mu\text{T})\hat{k},$$
and
$$\vec{B}(9.0\,\text{cm}) = -(133\,\mu\text{T})\hat{k} - (26.7\,\mu\text{T})\hat{k}$$

$$= \boxed{-(160\,\mu\text{T})\hat{k}}$$

33 •• As a student technician, you are preparing a lecture demonstration on "magnetic suspension. " You have a 16-cm long straight rigid wire that will be suspended by flexible conductive lightweight leads above a long straight wire. Currents that are equal but are in opposite directions will be established in the two wires so the 16-cm wire "floats" a distance h above the long wire with no tension in its suspension leads. If the mass of the 16-cm wire is 14 g and if h (the distance between the central axes of the two wires) is 1.5 mm, what should their common current be?

Picture the Problem The forces acting on the wire are the upward magnetic force F_B and the downward gravitational force mg, where m is the mass of the wire. We can use a condition for translational equilibrium and the expression for the force per unit length between parallel current-carrying wires to relate the required current to the mass of the wire, its length, and the separation of the two wires.

Apply $\sum F_y = 0$ to the floating wire to obtain:

$$F_B - mg = 0$$

Express the repulsive force acting on the upper wire:

$$F_B = 2\frac{\mu_0}{4\pi}\frac{I^2\ell}{R}$$

Substitute to obtain:

$$2\frac{\mu_0}{4\pi}\frac{I^2\ell}{R} - mg = 0 \Rightarrow I = \sqrt{\frac{4\pi mgR}{2\mu_0\ell}}$$

Substitute numerical values and evaluate I:

$$I = \sqrt{\frac{\left(14\times10^{-3}\,\text{kg}\right)\left(9.81\,\text{m/s}^2\right)\left(1.5\times10^{-3}\,\text{m}\right)}{2\left(10^{-7}\,\text{T}\cdot\text{m/A}\right)\left(0.16\,\text{m}\right)}} = \boxed{80\,\text{A}}$$

37 •• An infinitely long wire lies along the z axis and carries a current of 20 A in the $+z$ direction. A second infinitely long wire that is parallel to the z and intersects the x axis at $x = 10.0$ cm. (*a*) Find the current in the second wire if the magnetic field is zero at (2.0 cm, 0, 0) is zero. (*b*) What is the magnetic field at (5.0 cm, 0, 0)?

Picture the Problem Let the numeral 1 denote the current flowing along the positive z axis and the magnetic field resulting from it and the numeral 2 denote the current flowing in the wire located at $x = 10$ cm and the magnetic field resulting from it. We can express the magnetic field anywhere in the xy plane using $B = \frac{\mu_0}{4\pi}\frac{2I}{R}$ and the right-hand rule and then impose the condition that

$\vec{B} = 0$ to determine the current that satisfies this condition.

(a) Express the resultant magnetic field due to the two current-carrying wires:

$$\vec{B} = \vec{B}_1 + \vec{B}_2 \qquad (1)$$

Express the magnetic field at $x = 2.0$ cm due to the current flowing in the positive z direction:

$$\vec{B}_1(2.0\,\text{cm}) = \frac{\mu_0}{4\pi}\left(\frac{2I_1}{2.0\,\text{cm}}\right)\hat{j}$$

Express the magnetic field at $x = 2.0$ cm due to the current flowing in the wire at $x = 10.0$ cm:

$$\vec{B}_2(2.0\,\text{cm}) = -\frac{\mu_0}{4\pi}\left(\frac{2I_2}{8.0\,\text{cm}}\right)\hat{j}$$

Substitute for $\vec{B}_1$ and $\vec{B}_2$ in equation (1) and simplify to obtain:

$$\vec{B} = \frac{\mu_0}{4\pi}\left(\frac{2I_1}{2.0\,\text{cm}}\right)\hat{j} - \frac{\mu_0}{4\pi}\left(\frac{2I_2}{8.0\,\text{cm}}\right)\hat{j}$$

$$= \left(\frac{\mu_0}{4\pi}\frac{2I_1}{2.0\,\text{cm}} - \frac{\mu_0}{4\pi}\frac{2I_2}{8.0\,\text{cm}}\right)\hat{j}$$

For $\vec{B} = 0$:

$$\frac{\mu_0}{4\pi}\left(\frac{2I_1}{2.0\,\text{cm}}\right) - \frac{\mu_0}{4\pi}\left(\frac{2I_2}{8.0\,\text{cm}}\right) = 0$$

or

$$\frac{I_1}{2.0} - \frac{I_2}{8.0} = 0 \Rightarrow I_2 = 4I_1$$

Substitute numerical values and evaluate I_2:

$$I_2 = 4(20\,\text{A}) = \boxed{80\,\text{A}}$$

(b) Express the magnetic field at $x = 5.0$ cm:

$$\vec{B}(5.0\,\text{cm}) = \frac{\mu_0}{4\pi}\left(\frac{2I_1}{5.0\,\text{cm}}\right)\hat{j} - \frac{\mu_0}{4\pi}\left(\frac{2I_2}{5.0\,\text{cm}}\right)\hat{j} = \frac{2\mu_0}{4\pi(5.0\,\text{cm})}(I_1 - I_2)\hat{j}$$

Substitute numerical values and evaluate $\vec{B}(5.0\,\text{cm})$:

$$\vec{B}(5.0\,\text{cm}) = \frac{2(10^{-7}\,\text{T}\cdot\text{m/A})}{5.0\,\text{cm}}(20\,\text{A} - 80\,\text{A})\hat{j} = \boxed{-(0.24\,\text{mT})\hat{j}}$$

39 ••   Four long, straight parallel wires each carry current I. In a plane perpendicular to the wires, the wires are at the corners of a square of side length a. Find the magnitude of the force per unit length on one of the wires if (*a*) all the currents are in the same direction and (*b*) the currents in the wires at adjacent corners are oppositely directed.

Picture the Problem Choose a coordinate system with its origin at the lower left-hand corner of the square, the positive x axis to the right and the positive y axis upward. Let the numeral 1 denote the wire and current in the upper left-hand corner of the square, the numeral 2 the wire and current in the lower left-hand corner (at the origin) of the square, and the numeral 3 the wire and current in the lower right-hand corner of the square. We can use $B = \dfrac{\mu_0}{4\pi}\dfrac{2I}{R}$ and the right-hand rule to find the magnitude and direction of the magnetic field at, say, the upper right-hand corner due to each of the currents, superimpose these fields to find the resultant field, and then use $F = I\ell B$ to find the force per unit length on the wire.

(*a*) Express the resultant magnetic field at the upper right-hand corner:

$$\vec{B} = \vec{B}_1 + \vec{B}_2 + \vec{B}_3 \qquad\qquad (1)$$

When all the currents are into the paper their magnetic fields at the upper right-hand corner are as shown to the right:

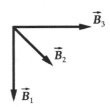

Express the magnetic field due to the current I_1:

$$\vec{B}_1 = -\frac{\mu_0}{4\pi}\frac{2I}{a}\,\hat{j}$$

Express the magnetic field due to the current I_2:

$$\vec{B}_2 = \frac{\mu_0}{4\pi}\frac{2I}{a\sqrt{2}}\cos 45^\circ\!\left(\hat{i} - \hat{j}\right)$$

$$= \frac{\mu_0}{4\pi}\frac{2I}{2a}\left(\hat{i} - \hat{j}\right)$$

Express the magnetic field due to the current I_3:

$$\vec{B}_3 = \frac{\mu_0}{4\pi}\frac{2I}{a}\,\hat{i}$$

Substitute in equation (1) and simplify to obtain:

$$\vec{B} = -\frac{\mu_0}{4\pi}\frac{2I}{a}\,\hat{j} + \frac{\mu_0}{4\pi}\frac{2I}{2a}\left(\hat{i} - \hat{j}\right) \ + \frac{\mu_0}{4\pi}\frac{2I}{a}\,\hat{i} = \frac{\mu_0}{4\pi}\frac{2I}{a}\left(-\hat{j} + \frac{1}{2}\left(\hat{i} - \hat{j}\right) + \hat{i}\right)$$

$$= \frac{\mu_0}{4\pi}\frac{2I}{a}\left[\left(1 + \frac{1}{2}\right)\hat{i} + \left(-1 - \frac{1}{2}\right)\hat{j}\right] = \frac{3\mu_0 I}{4\pi a}\left[\hat{i} - \hat{j}\right]$$

Using the expression for the magnetic force on a current-carrying wire, express the force per unit length on the wire at the upper right-hand corner:

$$\frac{F}{\ell} = BI \qquad\qquad (2)$$

Substitute to obtain:

$$\frac{\vec{F}}{\ell} = \frac{3\mu_0 I^2}{4\pi a}\left[\hat{i} - \hat{j}\right]$$

and

$$\frac{F}{\ell} = \sqrt{\left(\frac{3\mu_0 I^2}{4\pi a}\right)^2 + \left(\frac{3\mu_0 I^2}{4\pi a}\right)^2}$$

$$= \boxed{\frac{3\sqrt{2}\mu_0 I^2}{4\pi a}}$$

(b) When the current in the upper right-hand corner of the square is out of the page, and the currents in the wires at adjacent corners are oppositely directed, the magnetic fields at the upper right-hand are as shown to the right:

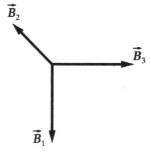

Express the magnetic field at the upper right-hand corner due to the current I_2:

$$\vec{B}_2 = \frac{\mu_0}{4\pi}\frac{2I}{a\sqrt{2}}\cos 45°\left(-\hat{i} + \hat{j}\right)$$

$$= \frac{\mu_0}{4\pi}\frac{2I}{2a}\left(-\hat{i} + \hat{j}\right)$$

Using $\vec{B}_1$ and $\vec{B}_3$ from (a), substitute in equation (1) and simplify to obtain:

$$\vec{B} = -\frac{\mu_0}{4\pi}\frac{2I}{a}\hat{j} + \frac{\mu_0}{4\pi}\frac{2I}{2a}\left(-\hat{i}+\hat{j}\right) + \frac{\mu_0}{4\pi}\frac{2I}{a}\hat{i} = \frac{\mu_0}{4\pi}\frac{2I}{a}\left(-\hat{j}+\frac{1}{2}\left(-\hat{i}+\hat{j}\right)+\hat{i}\right)$$

$$= \frac{\mu_0}{4\pi}\frac{2I}{a}\left[\left(1-\frac{1}{2}\right)\hat{i} + \left(-1+\frac{1}{2}\right)\hat{j}\right] = \frac{\mu_0}{4\pi}\frac{2I}{a}\left[\frac{1}{2}\hat{i} - \frac{1}{2}\hat{j}\right] = \frac{\mu_0 I}{4\pi a}\left[\hat{i}-\hat{j}\right]$$

Substitute in equation (2) to obtain: $$\frac{\vec{F}}{\ell} = \frac{\mu_0 I^2}{4\pi a}\left[\hat{i}-\hat{j}\right]$$

and

$$\frac{F}{\ell} = \sqrt{\left(\frac{\mu_0 I^2}{4\pi a}\right)^2 + \left(\frac{\mu_0 I^2}{4\pi a}\right)^2} = \boxed{\frac{\sqrt{2}\mu_0 I^2}{4\pi a}}$$

Magnetic Field Due to a Current-carrying Solenoid

41 ••   A solenoid that has length 30 cm, radius 1.2 cm, and 300 turns carries a current of 2.6 A. Find the magnitude of the magnetic field on the axis of the solenoid (a) at the center of the solenoid, (b) at one end of the solenoid.

Picture the Problem We can use $B_x = \frac{1}{2}\mu_0 nI\left(\dfrac{b}{\sqrt{b^2+R^2}} + \dfrac{a}{\sqrt{a^2+R^2}}\right)$ to find B at any point on the axis of the solenoid. Note that the number of turns per unit length for this solenoid is 300 turns/0.30 m = 1000 turns/m.

Express the magnetic field at any point on the axis of the solenoid: $$B_x = \tfrac{1}{2}\mu_0 nI\left(\frac{b}{\sqrt{b^2+R^2}} + \frac{a}{\sqrt{a^2+R^2}}\right)$$

Substitute numerical values to obtain:

$$B_x = \tfrac{1}{2}\left(4\pi\times10^{-7}\,\text{T}\cdot\text{m/A}\right)(1000)(2.6\,\text{A})\left(\frac{b}{\sqrt{b^2+(0.012\,\text{m})^2}} + \frac{a}{\sqrt{a^2+(0.012\,\text{m})^2}}\right)$$

$$= (1.634\,\text{mT})\left(\frac{b}{\sqrt{b^2+(0.012\,\text{m})^2}} + \frac{a}{\sqrt{a^2+(0.012\,\text{m})^2}}\right)$$

(a) Evaluate B_x for $a = b = 0.15$ m:

$$B_x(0.15\,\text{m}) = (1.634\,\text{mT})\left(\frac{0.15\,\text{m}}{\sqrt{(0.15\,\text{m})^2 + (0.012\,\text{m})^2}} + \frac{0.15\,\text{m}}{\sqrt{(0.15\,\text{m})^2 + (0.012\,\text{m})^2}}\right)$$

$$= \boxed{3.3\,\text{mT}}$$

(b) Evaluate B_x ($= B_{\text{end}}$) for $a = 0$ and $b = 0.30$ m:

$$B_x(0.30\,\text{m}) = (1.634\,\text{mT})\left(\frac{0.30\,\text{m}}{\sqrt{(0.30\,\text{m})^2 + (0.012\,\text{m})^2}}\right) = \boxed{1.6\,\text{mT}}$$

Note that $B_{\text{end}} \approx \frac{1}{2} B_{\text{center}}$.

Using Ampère's Law

45 •   A long, straight, thin-walled cylindrical shell of radius R carries a current I parallel to the central axis of the shell. Find the magnetic field (including direction) both inside and outside the shell.

Picture the Problem We can apply Ampère's law to a circle centered on the axis of the cylinder and evaluate this expression for $r < R$ and $r > R$ to find B inside and outside the cylinder. We can use the right-hand rule to determine the direction of the magnetic fields.

Apply Ampère's law to a circle centered on the axis of the cylinder:

$$\oint_C \vec{B} \cdot d\vec{\ell} = \mu_0 I_C$$

Note that, by symmetry, the field is the same everywhere on this circle.

Evaluate this expression for $r < R$:

$$\oint_C \vec{B}_{\text{inside}} \cdot d\vec{\ell} = \mu_0(0) = 0$$

Solve for B_{inside} to obtain:

$$B_{\text{inside}} = \boxed{0}$$

Evaluate this expression for $r > R$:

$$\oint_C \vec{B}_{\text{outside}} \cdot d\vec{\ell} = B(2\pi R) = \mu_0 I$$

Solve for B_{outside} to obtain:

$$B_{\text{outside}} = \boxed{\frac{\mu_0 I}{2\pi R}}$$

The direction of the magnetic field is in the direction of the curled fingers of your right hand when you grab the cylinder with your right thumb in the direction of the current.

47 ••   Show that a uniform magnetic field that has no fringing field, such as that shown in Figure 27- 56 is impossible because it violates Ampère's law. Do this calculation by applying Ampère's law to the rectangular curve shown by the dashed lines.

Determine the Concept The contour integral consists of four portions, two horizontal portions for which $\oint_C \vec{B} \cdot d\vec{\ell} = 0$, and two vertical portions. The portion within the magnetic field gives a nonvanishing contribution, whereas the portion outside the field gives no contribution to the contour integral. Hence, the contour integral has a finite value. However, it encloses no current; thus, it appears that Ampère's law is violated. What this demonstrates is that there must be a fringing field so that the contour integral does vanish.

49 ••   A long cylindrical shell has an inner radius a and an outer radius b and carries a current I parallel to the central axis. Assume that within the material of the shell the current density is uniformly distributed. Find an expression for the magnitude of the magnetic field for (a) $0 < R < a$, (b) $a < R < b$, and (c) $R > b$.

Picture the Problem We can use Ampère's law to calculate B because of the high degree of symmetry. The current through C depends on whether R is less than the inner radius a, greater than the inner radius a but less than the outer radius b, or greater than the outer radius b.

(a) Apply Ampère's law to a circular path of radius $R < a$ to obtain:

$$\oint_C \vec{B}_{r<a} \cdot d\vec{\ell} = \mu_0 I_C = \mu_0(0) = 0$$

and $B_{r<a} = \boxed{0}$

(b) Use the uniformity of the current over the cross-section of the conductor to express the current I' enclosed by a circular path whose radius satisfies the condition $a < R < b$:

$$\frac{I'}{\pi(R^2 - a^2)} = \frac{I}{\pi(b^2 - a^2)}$$

Solving for $I_C = I'$ yields:

$$I_C = I' = I\frac{R^2 - a^2}{b^2 - a^2}$$

Substitute in Ampère's law to obtain:

$$\oint_C \vec{B}_{a<R<b} \cdot d\vec{\ell} = B_{a<r<b}(2\pi R)$$

$$= \mu_0 I' = \mu_0 I \frac{R^2 - a^2}{b^2 - a^2}$$

Solving for $B_{a<r<b}$ yields:

$$\boxed{B_{a<r<b} = \frac{\mu_0 I}{2\pi r} \frac{R^2 - a^2}{b^2 - a^2}}$$

(c) Express I_C for $R > b$:

$$I_C = I$$

Substituting in Ampère's law yields:

$$\oint_C \vec{B}_{R>b} \cdot d\vec{\ell} = B_{r>b}(2\pi R) = \mu_0 I$$

Solve for $B_{R>b}$ to obtain:

$$\boxed{B_{R>b} = \frac{\mu_0 I}{2\pi R}}$$

51 ••   A tightly wound 1000-turn toroid has an inner radius 1.00 cm and an outer radius 2.00 cm, and carries a current of 1.50 A. The toroid is centered at the origin with the centers of the individual turns in the $z = 0$ plane. In the $z = 0$ plane: (a) What is the magnetic field strength at a distance of 1.10 cm from the origin? (b) What is the magnetic field strength at a distance of 1.50 cm from the origin?

Picture the Problem The magnetic field inside a tightly wound toroid is given by $B = \mu_0 NI/(2\pi r)$, where $a < r < b$ and a and b are the inner and outer radii of the toroid.

Express the magnetic field of a toroid:

$$B = \frac{\mu_0 NI}{2\pi r}$$

(a) Substitute numerical values and evaluate $B(1.10 \text{ cm})$:

$$B(1.10\,\text{cm}) = \frac{(4\pi \times 10^{-7}\,\text{N/A}^2)(1000)(1.50\,\text{A})}{2\pi(1.10\,\text{cm})} = \boxed{27.3\,\text{mT}}$$

(b) Substitute numerical values and evaluate $B(1.50 \text{ cm})$:

$$B(1.50\,\text{cm}) = \frac{(4\pi \times 10^{-7}\,\text{N/A}^2)(1000)(1.50\,\text{A})}{2\pi(1.50\,\text{cm})} = \boxed{20.0\,\text{mT}}$$

Magnetization and Magnetic Susceptibility

53 •   A tightly wound solenoid is 20.0-cm long, has 400 turns, and carries a current of 4.00 A so that its axial field is in the +z direction. Find B and B_{app} at the center when (a) there is no core in the solenoid, and (b) there is a soft iron core that has a magnetization of 1.2×10^6 A/m.

Picture the Problem We can use $B = B_{app} = \mu_0 n I$ to find B and B_{app} at the center when there is no core in the solenoid and $B = B_{app} + \mu_0 M$ when there is an iron core with a magnetization $M = 1.2 \times 10^6$ A/m.

(a) Express the magnetic field, in the absence of a core, in the solenoid :

$$B = B_{app} = \mu_0 n I$$

Substitute numerical values and evaluate B and B_{app}:

$$B = B_{app} = \left(4\pi \times 10^{-7} \text{ N/A}^2\right)\left(\frac{400}{0.200\,\text{m}}\right)(4.00\,\text{A}) = \boxed{10.1\,\text{mT}}$$

(b) With an iron core with a magnetization $M = 1.2 \times 10^6$ A/m present:

$$B_{app} = \boxed{10.1\,\text{mT}}$$

and

$$B = B_{app} + \mu_0 M = 10.1\,\text{mT} + \left(4\pi \times 10^{-7} \text{ N/A}^2\right)\left(1.2 \times 10^6 \text{ A/m}\right) = \boxed{1.5\,\text{T}}$$

57 ••   A cylinder of iron, initially unmagnetized, is cooled to 4.00 K. What is the magnetization of the cylinder at that temperature due to the influence of Earth's magnetic field of 0.300 G? Assume a magnetic moment of 2.00 Bohr magnetons per atom.

Picture the Problem We can use Curie's law to relate the magnetization M of the cylinder to its saturation magnetization M_S. The saturation magnetization is the product of the number of atoms n in the cylinder and the magnetic moment of each molecule. We can find n using the proportion $\dfrac{n}{N_A} = \dfrac{\rho_{Fe}}{M_{Fe}}$ where M_{Fe} is the molar mass of iron.

The magnetization of the cylinder is given by Curie's law:

$$M = \frac{1}{3}\frac{\mu B_{app}}{kT} M_S$$

Assuming a magnetic moment of 2.00 Bohr magnetons per atom:

$$M = \frac{1}{3}\frac{(2.00\mu_B)B_{Earth}}{kT}M_S \qquad (1)$$

The saturation magnetization is given by:

$$M_S = n\mu = n(2.00\mu_B)$$

where n is the number of atoms and μ is the magnetic moment of each molecule.

The number of atoms of iron per unit volume n can be found from the molar mass M_{Fe} of iron, the density ρ_{Fe} of iron, and Avogadro's number N_A:

$$n = \frac{\rho_{Fe}}{M_{Fe}}N_A \Rightarrow M_S = \frac{2.00\rho_{Fe}N_A\mu_B}{M_{Fe}}$$

Substitute numerical values and evaluate M_S:

$$M_S = \frac{2.00\left(7.96\times10^3\,\frac{kg}{m^3}\right)\left(6.022\times10^{23}\,\frac{atoms}{mol}\right)\left(9.27\times10^{-24}\,A\cdot m^2\right)}{55.85\,\frac{g}{mol}\times\frac{1\,kg}{10^3\,g}}$$

$$= 1.591\times10^6\,\frac{A}{m}$$

Substitute numerical values in equation (1) and evaluate M:

$$M = \frac{2.00\left(5.788\times10^{-5}\,\frac{eV}{T}\right)\left(0.300\,G\times\frac{1\,T}{10^4\,G}\right)}{3\left(8.617\times10^{-5}\,\frac{eV}{K}\right)(4.00\,K)}\left(1.591\times10^6\,\frac{A}{m}\right) = \boxed{5.34\,\frac{A}{m}}$$

Paramagnetism

65 ••   A toroid has N turns, carries a current I, has a mean radius R, and has a cross-sectional radius r, where $r << R$ (Figure 27-59). When the toroid is filled with material, it is called a *Rowland ring*. Find B_{app} and B in such a ring, assuming a magnetization that is everywhere parallel to $\vec{B}_{app}$.

Picture the Problem We can use $B_{app} = \frac{\mu_0 NI}{2\pi a}$ to express B_{app} and $B = B_{app} + \mu_0 M$ to express B in terms of B_{app} and M.

Express B_{app} inside a tightly wound toroid:

$$B_{app} = \boxed{\frac{\mu_0 NI}{2\pi a}} \text{ for } R - r < a < R + r$$

The resultant field B in the ring is the sum of B_{app} and $\mu_0 M$:

$$B = B_{app} + \mu_0 M = \boxed{\frac{\mu_0 NI}{2\pi a} + \mu_0 M}$$

Ferromagnetism

69 ••   The saturation magnetization for annealed iron occurs when $B_{app} = 0.201$ T. Find the permeability and the relative permeability of annealed iron at saturation. (See Table 27-2)

Picture the Problem We can relate the permeability μ of annealed iron to χ_m using $\mu = (1 + \chi_m)\mu_0$, find χ_m using $M = \chi_m \dfrac{B_{app}}{\mu_0}$, and use its definition

$(K_m = 1 + \chi_m)$ to evaluate K_m.

Express the permeability μ of annealed iron in terms of its magnetic susceptibility χ_m:

$$\mu = (1 + \chi_m)\mu_0 \qquad (1)$$

The magnetization M in terms of χ_m and B_{app} is given by:

$$M = \chi_m \frac{B_{app}}{\mu_0}$$

Solve for and evaluate χ_m (see Table 27-2 for the product of μ_0 and M):

$$\chi_m = \frac{\mu_0 M}{B_{app}} = \frac{2.16 \text{ T}}{0.201 \text{ T}} = 10.75$$

Use its definition to express and evaluate the relative permeability K_m:

$$K_m = 1 + \chi_m = 1 + 10.75 = 11.746$$
$$= \boxed{11.7}$$

Substitute numerical values in equation (1) and evaluate μ:

$$\mu = (1 + 10.746)(4\pi \times 10^{-7} \text{ N/A}^2)$$
$$= \boxed{1.48 \times 10^{-5} \text{ N/A}^2}$$

73 ••   A toroid has N turns, carries a current I, has a mean radius R, and has a cross-sectional radius r, where $r \ll R$ (Figure 27-53). The core of the toroid of is filled with iron. When the current is 10.0 A, the magnetic field in the region where the iron is has a magnitude of 1.80 T. (*a*) What is the magnetization? (*b*) Find the values for the relative permeability, the permeability, and magnetic susceptibility for this iron sample.

Picture the Problem We can use $B = B_{app} + \mu_0 M$ and the expression for the magnetic field inside a tightly wound toroid to find the magnetization M. We can find K_m from its definition, $\mu = K_m \mu_0$ to find μ, and $K_m = 1 + \chi_m$ to find χ_m for the iron sample.

(a) Relate the magnetization to B and B_{app}:

$$B = B_{app} + \mu_0 M \Rightarrow M = \frac{B - B_{app}}{\mu_0}$$

Express the magnetic field inside a tightly wound toroid:

$$B_{app} = \frac{\mu_0 NI}{2\pi r}$$

Substitute for B_{app} and simplify to obtain:

$$M = \frac{B - \frac{\mu_0 NI}{2\pi r}}{\mu_0} = \frac{B}{\mu_0} - \frac{NI}{2\pi r}$$

Substitute numerical values and evaluate M:

$$M = \frac{1.80\,\text{T}}{4\pi \times 10^{-7}\,\text{N/A}^2} - \frac{2000(10.0\,\text{A})}{2\pi(0.200\,\text{m})}$$

$$= \boxed{1.42 \times 10^6\,\text{A/m}}$$

(b) Use its definition to express K_m:

$$K_m = \frac{B}{B_{app}} = \frac{B}{\frac{\mu_0 NI}{2\pi r}} = \frac{2\pi r B}{\mu_0 NI}$$

Substitute numerical values and evaluate K_m:

$$K_m = \frac{2\pi(0.200\,\text{m})(1.80\,\text{T})}{(4\pi \times 10^{-7}\,\text{N/A}^2)(2000)(10.0\,\text{A})}$$

$$= \boxed{90.0}$$

Now that we know K_m we can find μ using:

$$\mu = K_m \mu_0 = 90(4\pi \times 10^{-7}\,\text{N/A}^2)$$

$$= \boxed{1.13 \times 10^{-4}\,\text{T} \cdot \text{m/A}}$$

Relate χ_m to K_m:

$$K_m = 1 + \chi_m \Rightarrow \chi_m = K_m - 1$$

Substitute the numerical value of K_m and evaluate χ_m:

$$\chi_m = 90 - 1 = \boxed{89}$$

General Problems

77 •   Using Figure 27-61, find the magnetic field (in terms of the parameters given in the figure) at point P, the common center of the two arcs.

Picture the Problem Let out of the page be the positive x direction. Because point P is on the line connecting the straight segments of the conductor, these segments do not contribute to the magnetic field at P. Hence, the resultant magnetic field at P will be the sum of the magnetic fields due to the current in the two semicircles, and we can use the expression for the magnetic field at the center of a current loop to find $\vec{B}_P$.

Express the resultant magnetic field at P:	$\vec{B}_P = \vec{B}_1 + \vec{B}_2$	(1)

Express the magnetic field at the center of a current loop:

$$B = \frac{\mu_0 I}{2R}$$

where R is the radius of the loop.

Express the magnetic field at the center of half a current loop:

$$B = \frac{1}{2}\frac{\mu_0 I}{2R} = \frac{\mu_0 I}{4R}$$

Express $\vec{B}_1$ and $\vec{B}_2$:

$$\vec{B}_1 = \frac{\mu_0 I}{4R_1}\hat{i} \text{ and } \vec{B}_2 = -\frac{\mu_0 I}{4R_2}\hat{i}$$

Substitute in equation (1) to obtain:

$$\vec{B}_P = \frac{\mu_0 I}{4R_1}\hat{i} - \frac{\mu_0 I}{4R_2}\hat{i}$$

$$\boxed{= \frac{\mu_0 I}{4}\left(\frac{1}{R_1} - \frac{1}{R_2}\right)\hat{i} \text{ out of the page}}$$

81 •• A long straight wire carries a current of 20.0 A, as shown in Figure 27-63. A rectangular coil that has two sides parallel to the straight wire has sides that are 5.00-cm long and 10.0-cm long. The side nearest to the wire is 2.00 cm from the wire. The coil carries a current of 5.00 A. (*a*) Find the force on each segment of the rectangular coil due to the current in the long straight wire. (*b*) What is the net force on the coil?

Picture the Problem Let I_1 and I_2 represent the currents of 20 A and 5.0 A, $\vec{F}_{\text{top}}$, $\vec{F}_{\text{left side}}$, $\vec{F}_{\text{bottom}}$, and $\vec{F}_{\text{right side}}$ the forces that act on the horizontal wire, and $\vec{B}_1$, $\vec{B}_2$, $\vec{B}_3$, and $\vec{B}_4$ the magnetic fields at these wire segments due to I_1. We'll need to take into account the fact that $\vec{B}_1$ and $\vec{B}_3$ are not constant over the segments 1 and 3 of the rectangular coil. Let the $+x$ direction be to the right and the $+y$ direction be upward. Then the $+z$ direction is toward you (i.e., out of the page). Note that only the components of $\vec{B}_1$, $\vec{B}_2$, $\vec{B}_3$, and $\vec{B}_4$ into or out of the page contribute to the forces acting on the rectangular coil. The $+x$ and $+y$ directions are up the page and to the right.

(a) Express the force $d\vec{F}_1$ acting on a current element $I_2\,d\vec{\ell}$ in the top segment of wire:

$$d\vec{F}_{top} = I_2\,d\vec{\ell} \times \vec{B}_1$$

Because $I_2\,d\vec{\ell} = I_2\,d\ell\left(-\hat{i}\right)$ in this segment of the coil and the magnetic field due to I_1 is given by $\vec{B}_1 = \dfrac{\mu_0 I_1}{2\pi\,\ell}\left(-\hat{k}\right)$:

$$\vec{F}_{top} = I_2\,d\ell\left(-\hat{i}\right) \times \frac{\mu_0 I_1}{2\pi\,\ell}\left(-\hat{k}\right)$$

$$= -\frac{\mu_0 I_1 I_2}{2\pi}\frac{d\ell}{\ell}\,\hat{j}$$

Integrate $d\vec{F}_{top}$ to obtain:

$$\vec{F}_{top} = -\frac{\mu_0 I_1 I_2}{2\pi}\int_{2.0\,cm}^{7.0\,cm}\frac{d\ell}{\ell}\,\hat{j}$$

$$= -\frac{\mu_0 I_1 I_2}{2\pi}\ln\left(\frac{7.0\,cm}{2.0\,cm}\right)\hat{j}$$

Substitute numerical values and evaluate $\vec{F}_{top}$:

$$\vec{F}_{top} = -\frac{\left(4\pi\times10^{-7}\,\dfrac{N}{A^2}\right)(20\,A)(5.0\,A)}{2\pi}\ln\left(\frac{7.0\,cm}{2.0\,cm}\right)\hat{j} = \boxed{-\left(2.5\times10^{-5}\,N\right)\hat{j}}$$

Express the force $d\vec{F}_{bottom}$ acting on a current element $I_2\,d\vec{\ell}$ in the horizontal segment of wire at the bottom of the coil:

$$d\vec{F}_{bottom} = I_2\,d\vec{\ell} \times \vec{B}_3$$

Because $I_2\,d\vec{\ell} = I_2\,d\ell\left(\hat{i}\right)$ in this segment of the coil and the magnetic field due to I_1 is given by $\vec{B}_1 = \dfrac{\mu_0 I_1}{2\pi\,\ell}\left(-\hat{k}\right)$:

$$d\vec{F}_{bottom} = I_2\,d\ell\,\hat{i} \times \frac{\mu_0 I_1}{2\pi\ell}\left(-\hat{k}\right)$$

$$= \frac{\mu_0 I_1 I_2}{2\pi}\frac{d\ell}{\ell}\,\hat{j}$$

Integrate $d\vec{F}_{bottom}$ to obtain:

$$dF_{bottom} = \frac{\mu_0 I_1 I_2}{2\pi} \int_{2.0\,cm}^{7.0\,cm} \frac{d\ell}{\ell}\hat{j}$$

$$= \frac{\mu_0 I_1 I_2}{2\pi} \ln\left(\frac{7.0\,cm}{2.0\,cm}\right)\hat{j}$$

Substitute numerical values and evaluate $\vec{F}_{bottom}$:

$$\vec{F}_{bottom} = \frac{\left(4\pi \times 10^{-7} \dfrac{N}{A^2}\right)(20\,A)(5.0\,A)}{2\pi} \ln\left(\frac{7.0\,cm}{2.0\,cm}\right)\hat{j} = \boxed{\left(2.5 \times 10^{-5}\,N\right)\hat{j}}$$

Express the forces $\vec{F}_{left\,side}$ and $\vec{F}_{right\,side}$ in terms of I_2 and $\vec{B}_2$ and $\vec{B}_4$:

$$\vec{F}_{left\,side} = I_2\vec{\ell}_2 \times \vec{B}_2$$
and
$$\vec{F}_{right\,side} = I_2\vec{\ell}_4 \times \vec{B}_4$$

Express $\vec{B}_2$ and $\vec{B}_4$:

$$\vec{B}_2 = -\frac{\mu_0}{4\pi}\frac{2I_1}{R_1}\hat{k} \quad and \quad \vec{B}_4 = -\frac{\mu_0}{4\pi}\frac{2I_1}{R_4}\hat{k}$$

Substitute for $\vec{B}_2$ and $\vec{B}_4$ to obtain:

$$\vec{F}_{left\,side} = -I_2\ell_2\hat{j} \times \left(-\frac{\mu_0}{4\pi}\frac{2I_1}{R_1}\hat{k}\right) = \frac{\mu_0\ell_2 I_1 I_2}{2\pi R_2}\hat{i}$$

and

$$\vec{F}_{right\,side} = I_2\ell_4\hat{j} \times \left(-\frac{\mu_0}{4\pi}\frac{2I_1}{R_4}\hat{k}\right) = -\frac{\mu_0\ell_4 I_1 I_2}{2\pi R_4}\hat{i}$$

Substitute numerical values and evaluate $\vec{F}_{left\,side}$ and $\vec{F}_{right\,side}$:

$$\vec{F}_{left\,side} = \frac{\left(4\pi \times 10^{-7}\,N/A^2\right)(0.100\,m)(20.0\,A)(5.00\,A)}{2\pi(0.0200\,m)}\hat{i} = \boxed{\left(1.0 \times 10^{-4}\,N\right)\hat{i}}$$

and

$$\vec{F}_{right\,side} = -\frac{\left(4\pi \times 10^{-7}\,N/A^2\right)(0.100\,m)(20.0\,A)(5.00\,A)}{2\pi(0.0700\,m)}\hat{i} = \boxed{\left(-0.29 \times 10^{-4}\,N\right)\hat{i}}$$

(b) Express the net force acting on the coil:

$$\vec{F}_{net} = \vec{F}_{top} + \vec{F}_{left\,side} + \vec{F}_{bottom} + \vec{F}_{right\,side}$$

Substitute for $\vec{F}_{top}$, $\vec{F}_{left\,side}$, $\vec{F}_{bottom}$, and $\vec{F}_{right\,side}$ and simplify to obtain:

$$\vec{F}_{net} = \left(-2.5\times10^{-5}\ N\right)\hat{j} + \left(1.0\times10^{-4}\ N\right)\hat{i} + \left(2.5\times10^{-5}\ N\right)\hat{j} + \left(-0.29\times10^{-4}\ N\right)\hat{i}$$

$$= \boxed{\left(0.71\times10^{-4}\ N\right)\hat{i}}$$

95 ••   A current balance is constructed in the following way: A straight 10.0-cm-long section of wire is placed on top of the pan of an electronic balance (Figure 27-69). This section of wire is connected in series with a power supply and a long straight horizontal section of wire that is parallel to it and positioned directly above it. The distance between the central axes of the two wires is 2.00 cm. The power supply provides a current in the wires. When the power supply is switched on, the reading on the balance increases by 5.00 mg. What is the current in the wire?

Picture the Problem The force acting on the lower wire is given by $F_{lower\,wire} = I\ell B$, where I is the current in the lower wire, ℓ is the length of the wire on the balance, and B is the magnetic field strength at the location of the lower wire due to the current in the upper wire. We can apply Ampere's law to find B at the location of the wire on the pan of the balance.

The force experienced by the lower wire is given by:

$$F_{lower\,wire} = I\ell B$$

Apply Ampere's law to a closed circular path of radius r centered on the upper wire to obtain:

$$B(2\pi r) = \mu_0 I_C = \mu_0 I \Rightarrow B = \frac{\mu_0 I}{2\pi r}$$

Substituting for B in the expression for the force on the lower wire and simplifying yields:

$$F_{lower\,wire} = I\ell\left(\frac{\mu_0 I}{2\pi r}\right) = \frac{\mu_0 \ell I^2}{2\pi r}$$

Solve for I to obtain:

$$I = \sqrt{\frac{2\pi r F_{lower\,wire}}{\mu_0 \ell}}$$

Note that the force on the lower wire is the increase in the reading of the balance. Substitute numerical values and evaluate I:

$$I = \sqrt{\frac{2\pi(2.00\ cm)(5.00\times10^{-6}\ kg)}{(4\pi\times10^{-7}\ N/A^2)(10.0\ cm)}}$$

$$= 2.236\ A = \boxed{2.24\ A}$$

97 •••   A non-conducting disk that has radius R, carries a uniform surface charge density σ, and rotates with angular speed ω. (a) Consider an annular strip that has a radius r, a width dr, and a charge dq. Show that the current (dI) produced by this rotating strip is given by $\omega\sigma r\,dr$. (b) Use your result from

Part (*a*) to show that the magnetic field strength at the center of the disk is given by the expression $\frac{1}{2}\mu_0\sigma\omega R$. (*c*) Use your result from Part (*a*) to find an expression for the magnetic field strength at a point on the central axis of the disk a distance *z* from its center.

Picture the Problem The diagram shows the rotating disk and the circular strip of radius *r* and width *dr* with charge *dq*. We can use the definition of surface charge density to express *dq* in terms of *r* and *dr* and the definition of current to show that $dI = \omega\sigma r\,dr$. We can then use this current and expression for the magnetic field on the axis of a current loop to obtain the results called for in Parts (*b*) and (*c*).

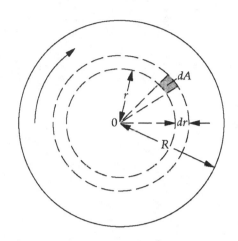

(*a*) Express the total charge *dq* that passes a given point on the circular strip once each period:

$$dq = \sigma dA = 2\pi\sigma r\,dr$$

Letting *q* be the total charge that passes along a radial section of the disk in a period of time *T*, express the current in the element of width *dr*:

$$dI = \frac{dq}{dt} = \frac{2\pi\sigma r\,dr}{\dfrac{2\pi}{\omega}} = \boxed{\omega\sigma r\,dr}$$

(*c*) Express the magnetic field dB_x at a distance *z* along the axis of the disk due to the current loop of radius *r* and width *dr*:

$$dB_x = \frac{\mu_0}{4\pi}\frac{2\pi r^2\,dI}{\left(z^2+r^2\right)^{3/2}}$$

$$= \frac{\mu_0\omega\sigma r^3}{2\left(z^2+r^2\right)^{3/2}}\,dr$$

Integrate from $r = 0$ to $r = R$ to obtain:

$$B_x = \frac{\mu_0\omega\sigma}{2}\int_0^R \frac{r^3}{\left(z^2+r^2\right)^{3/2}}\,dr$$

$$= \boxed{\frac{\mu_0\omega\sigma}{2}\left(\frac{R^2+2z^2}{\sqrt{R^2+z^2}}-2x\right)}$$

(*b*) Evaluate B_x for $x = 0$:

$$B_x(0) = \frac{\mu_0\omega\sigma}{2}\left(\frac{R^2}{\sqrt{R^2}}\right) = \boxed{\tfrac{1}{2}\mu_0\sigma\omega R}$$

Chapter 28
Magnetic Induction

Conceptual Problems

1 •   (a) The magnetic equator is a line on the surface of Earth on which Earth's magnetic field is horizontal. At the magnetic equator, how would you orient a flat sheet of paper so as to create the maximum magnitude of magnetic flux through it? (b) How about the minimum magnitude of magnetic flux?

Determine the Concept
(a) Orient the sheet so the normal to the sheet is both horizontal and perpendicular to the local tangent to the magnetic equator.

(b) Orient the sheet of paper so the normal to the sheet is perpendicular to the direction of the normal described in the answer to Part (a).

3 •   Show that the following combination of SI units is equivalent to the volt: $T \cdot m^2/s$.

Determine the Concept Because a volt is a joule per coulomb, we can show that the SI units $\dfrac{T \cdot m^2}{s}$ are equivalent to a volt by making a series of substitutions and simplifications that reduces these units to a joule per coulomb.

The units of a tesla are $\dfrac{N}{A \cdot m}$:

$$\frac{T \cdot m^2}{s} = \frac{\dfrac{N}{A \cdot m} \cdot m^2}{s} = \frac{\dfrac{N \cdot m}{A}}{s}$$

Substitute the units of an ampere (C/s), replace $N \cdot m$ with J, and simplify to obtain:

$$\frac{T \cdot m^2}{s} = \frac{\dfrac{J}{\dfrac{C}{s}}}{s} = \frac{J}{C}$$

Finally, because a joule per coulomb is a volt:

$$\frac{T \cdot m^2}{s} = \boxed{V}$$

5 •   A current is induced in a conducting loop that lies in a horizontal plane and the induced current is clockwise when viewed from above. Which of the following statements could be true? (a) A constant magnetic field is directed vertically downward. (b) A constant magnetic field is directed vertically upward. (c) A magnetic field whose magnitude is increasing is directed vertically downward. (*d*) A magnetic field whose magnitude is decreasing is directed vertically downward. (*e*) A magnetic field whose magnitude is decreasing is directed vertically upward.

Determine the Concept We know that the magnetic flux (in this case the magnetic field because the area of the conducting loop is constant and its orientation is fixed) must be changing so the only issues are whether the field is increasing or decreasing and in which direction. Because the direction of the magnetic field associated with the clockwise current is vertically downward, the changing field that is responsible for it must be either increasing vertically upward (not included in the list of possible answers) or a decreasing field directed into the page. $\boxed{(d)}$ is correct.

7 •   The planes of the two circular loops in Figure 28-38, are parallel. As viewed from the left, a counterclockwise current exists in loop A. If the magnitude of the current in loop A is increasing, what is the direction of the current induced in loop B? Do the loops attract or repel each other? Explain your answer.

Determine the Concept Clockwise as viewed from the left. The loops repel each other.

15 •   True or false:

(*a*) The induced emf in a circuit is equal to the negative of the magnetic flux through the circuit.
(*b*) There can be a non-zero induced emf at an instant when the flux through the circuit is equal to zero.
(*c*) The self inductance of a solenoid is proportional to the rate of change of the current in the solenoid.
(*d*) The magnetic energy density at some point in space is proportional to the square of the magnitude of the magnetic field at that point.
(*e*) The inductance of a solenoid is proportional to the current in it.

(*a*) False. The induced emf in a circuit is equal to *the rate of change of* the magnetic flux through the circuit.

(*b*) True. The rate of change of the magnetic flux can be non-zero when the flux through the circuit is momentarily zero

(*c*) False. The self inductance of a solenoid is determined by its length, cross-sectional area, number of turns per unit length, and the permeability of the matter in its core.

(d) True. The magnetic energy density at some point in space is given by Equation 28-20: $u_m = \dfrac{B^2}{2\mu_0}$.

(e) False. The inductance of a solenoid is determined by its length, cross-sectional area, number of turns per unit length, and the permeability of the matter in its core.

Magnetic Flux

21 •   A circular coil has 25 turns and a radius of 5.0 cm. It is at the equator, where Earth's magnetic field is 0.70 G, north. The axis of the coil is the line that passes through the center of the coil and is perpendicular to the plane of the coil. Find the magnetic flux through the coil when the axis of the coil is (a) vertical, (b) horizontal with the axis pointing north, (c) horizontal with the axis pointing east, and (d) horizontal with the axis making an angle of 30° with north.

Picture the Problem Because the coil defines a plane with area A and $\vec{B}$ is constant in magnitude and direction over the surface and makes an angle θ with the unit normal vector, we can use $\phi_m = NBA\cos\theta$ to find the magnetic flux through the coil.

The magnetic flux through the coil is given by:

$$\phi_m = NBA\cos\theta = NB\pi r^2 \cos\theta$$

Substitute for numerical values to obtain:

$$\phi_m = 25\left(0.70\,\mathrm{G}\cdot\frac{1\,\mathrm{T}}{10^4\,\mathrm{G}}\right)\pi\left(5.0\times10^{-2}\,\mathrm{m}\right)^2\cos\theta = (13.7\,\mu\mathrm{Wb})\cos\theta$$

(a) When the plane of the coil is horizontal, $\theta = 90°$:

$$\phi_m = (13.7\,\mu\mathrm{Wb})\cos90° = \boxed{0}$$

(b) When the plane of the coil is vertical with its axis pointing north, $\theta = 0°$:

$$\phi_m = (13.7\,\mu\mathrm{Wb})\cos0° = \boxed{14\,\mu\mathrm{Wb}}$$

(c) When the plane of the coil is vertical with its axis pointing east, $\theta = 90°$:

$$\phi_m = (13.7\,\mu\mathrm{Wb})\cos90° = \boxed{0}$$

(d) When the plane of the coil is vertical with its axis making an angle of 30° with north, $\theta = 30°$:

$$\phi_m = (13.7\,\mu\text{Wb})\cos 30° = \boxed{12\,\mu\text{Wb}}$$

27 ••   A long solenoid has n turns per unit length, has a radius R_1, and carries a current I. A circular coil with radius R_2 and with N total turns is coaxial with the solenoid and equidistant from its ends. (a) Find the magnetic flux through the coil if $R_2 > R_1$. (b) Find the magnetic flux through the coil if $R_2 < R_1$.

Picture the Problem The magnetic field outside the solenoid is, to a good approximation, zero. Hence, the flux through the coil is the flux in the core of the solenoid. The magnetic field inside the solenoid is uniform. Hence, the flux through the circular coil is given by the same expression with R_2 replacing R_1:

(a) The flux through the large circular loop outside the solenoid is given by:

$$\phi_m = NBA$$

Substituting for B and A and simplifying yields:

$$\phi_m = N(\mu_0 nI)(\pi R_1^2) = \boxed{\mu_0 nIN\pi R_1^2}$$

(b) The flux through the coil when $R_2 < R_1$ is given by:

$$\phi_m = N(\mu_0 nI)(\pi R_2^2) = \boxed{\mu_0 nIN\pi R_2^2}$$

Induced EMF and Faraday's Law

33 ••   A 100-turn circular coil has a diameter of 2.00 cm, a resistance of 50.0 Ω, and the two ends of the coil are connected together. The plane of the coil is perpendicular to a uniform magnetic field of magnitude 1.00 T. The direction of the field is reversed. (a) Find the total charge that passes through a cross section of the wire. If the reversal takes 0.100 s, find (b) the average current and (c) the average emf during the reversal.

Picture the Problem We can use the definition of average current to express the total charge passing through the coil as a function of I_{av}. Because the induced current is proportional to the induced emf and the induced emf, in turn, is given by Faraday's law, we can express ΔQ as a function of the number of turns of the coil, the magnetic field, the resistance of the coil, and the area of the coil. Knowing the reversal time, we can find the average current from its definition and the average emf from Ohm's law.

(a) Express the total charge that passes through the coil in terms of the induced current:

$$\Delta Q = I_{av}\Delta t \qquad (1)$$

Relate the induced current to the induced emf:

$$I = I_{av} = \frac{\mathcal{E}}{R}$$

Using Faraday's law, express the induced emf in terms of ϕ_m:

$$\mathcal{E} = -\frac{\Delta\phi_m}{\Delta t}$$

Substitute in equation (1) and simplify to obtain:

$$\Delta Q = \frac{|\mathcal{E}|}{R}\Delta t - \frac{\frac{\Delta\phi_m}{\Delta t}}{R}\Delta t - \frac{2\phi_m}{R}$$

$$= \frac{2NBA}{R} = \frac{2NB\left(\frac{\pi}{4}d^2\right)}{R}$$

$$= \frac{NB\pi d^2}{2R}$$

where d is the diameter of the coil.

Substitute numerical values and evaluate ΔQ:

$$\Delta Q = \frac{(100)(1.00\,\text{T})\pi(0.0200\,\text{m})^2}{2(50.0\,\Omega)}$$

$$= 1.257\,\text{mC} = \boxed{1.26\,\text{mC}}$$

(b) Apply the definition of average current to obtain:

$$I_{av} = \frac{\Delta Q}{\Delta t} = \frac{1.257\,\text{mC}}{0.100\,\text{s}} = 12.57\,\text{mA}$$

$$= \boxed{12.6\,\text{mA}}$$

(c) Using Ohm's law, relate the average emf in the coil to the average current:

$$\mathcal{E}_{av} = I_{av}R = (12.57\,\text{mA})(50.0\,\Omega)$$

$$= \boxed{628\,\text{mV}}$$

Motional EMF

41 •• In Figure 28-47, the rod has a mass m and a resistance R. The rails are horizontal, frictionless and have negligible resistances. The distance between the rails is ℓ. An ideal battery that has an emf $\mathcal{E}$ is connected between points a and b so that the current in the rod is downward. The rod released from rest at $t = 0$. (a) Derive an expression for the force on the rod as a function of the speed. (b) Show that the speed of the rod approaches a terminal speed and find an expression for the terminal speed. (*c*) What is the current when the rod is moving at its terminal speed?

Picture the Problem (*a*) The net force acting on the rod is the magnetic force it experiences as a consequence of carrying a current and being in a magnetic field. The net emf that drives *I* in this circuit is the difference between the emf of the battery and the emf induced in the rod as a result of its motion. Applying a right-hand rule to the rod reveals that the direction of this magnetic force is to the right. Hence the rod will accelerate to the right when it is released. (*b*) We can obtain the equation of motion of the rod by applying Newton's 2nd law to relate its acceleration to $\mathcal{E}, B, I, R$ and ℓ. (*c*) Letting $v = v_t$ in the equation for the current in the circuit will yield current when the rod is at its terminal speed.

(*a*) Express the magnetic force on the current-carrying rod:

$$F_m = I\ell B$$

The current in the rod is given by:

$$I = \frac{\mathcal{E} - B\ell v}{R} \tag{1}$$

Substituting for *I* yields:

$$F_m = \left(\frac{\mathcal{E} - B\ell v}{R}\right)\ell B = \boxed{\frac{B\ell}{R}(\mathcal{E} - B\ell v)}$$

(*b*) Letting the direction of motion of the rod be the positive *x* direction, apply $\sum F_x = ma_x$ to the rod:

$$\frac{B\ell}{R}(\mathcal{E} - B\ell v) = m\frac{dv}{dt}$$

Solving for dv/dt yields:

$$\frac{dv}{dt} = \frac{B\ell}{mR}(\mathcal{E} - B\ell v)$$

Note that as *v* increases, $\mathcal{E} - B\ell v \to 0$, $dv/dt \to 0$ and the rod approaches its terminal speed v_t. Set $dv/dt = 0$ to obtain:

$$\frac{B\ell}{mR}(\mathcal{E} - B\ell v_t) = 0 \Rightarrow v_t = \boxed{\frac{\mathcal{E}}{B\ell}}$$

(*c*) Substitute v_t for *v* in equation (1) to obtain:

$$I = \frac{\mathcal{E} - B\ell\dfrac{\mathcal{E}}{B\ell}}{R} = \boxed{0}$$

45 •   A 2.00-cm by 1.50-cm rectangular coil has 300 turns and rotates in a region that has a magnetic field of 0.400 T. (*a*) What is the maximum emf generated when the coil rotates at 60 rev/s? (*b*) What must its angular speed be to generate a maximum emf of 110 V?

Picture the Problem We can use the relationship $\mathcal{E}_{max} = 2\pi NBAf$ to relate the maximum emf generated to the area of the coil, the number of turns of the coil, the magnetic field in which the coil is rotating, and the angular speed at which it rotates.

(*a*) Relate the induced emf to the magnetic field in which the coil is rotating:

$$\mathcal{E}_{max} = NBA\omega = 2\pi NBAf \qquad (1)$$

Substitute numerical values and evaluate $\mathcal{E}_{max}$:

$$\mathcal{E}_{max} = 2\pi(300)(0.400\,\text{T})(2.00\times10^{-2}\,\text{m})(1.50\times10^{-2}\,\text{m})(60\,\text{s}^{-1}) = \boxed{14\,\text{V}}$$

(*b*) Solve equation (1) for f:

$$f = \frac{\mathcal{E}_{max}}{2\pi NBA}$$

Substitute numerical values and evaluate f:

$$f = \frac{110\,\text{V}}{2\pi(300)(0.400\,\text{T})(2.00\times10^{-2}\,\text{m})(1.50\times10^{-2}\,\text{m})} = \boxed{486\,\text{rev/s}}$$

Inductance

49 ••   An insulated wire that has a resistance of 18.0 Ω/m and a length of 9.00 m will be used to construct a resistor. First, the wire is bent in half and then the doubled wire is wound on a cylindrical form (Figure 28-50) to create a 25.0-cm-long helix that has a diameter equal to 2.00 cm. Find both the resistance and the inductance of this wire-wound resistor.

Picture the Problem Note that the current in the two parts of the wire is in opposite directions. Consequently, the total flux in the coil is zero. We can find the resistance of the wire-wound resistor from the length of wire used and the resistance per unit length.

Because the total flux in the coil is zero:

$$L = \boxed{0}$$

Express the total resistance of the wire:	$R = \left(18.0\dfrac{\Omega}{m}\right)L$

Substitute numerical values and evaluate R:	$R = \left(18.0\dfrac{\Omega}{m}\right)(9.00\,m) = \boxed{162\,\Omega}$

53 •••   Show that the inductance of a toroid of rectangular cross section, as shown in Figure 28-52 is given by $L = \dfrac{\mu_0 N^2 H \ln(b/a)}{2\pi}$ where N is the total number of turns, a is the inside radius, b is the outside radius, and H is the height of the toroid.

Picture the Problem We can use Ampere's law to express the magnetic field inside the rectangular toroid and the definition of magnetic flux to express ϕ_m through the toroid. We can then use the definition of self-inductance of a solenoid to express L.

Using the definition of the self-inductance of a solenoid, express L in terms of ϕ_m, N, and I:	$L = \dfrac{N\phi_m}{I}$	(1)

Apply Ampere's law to a closed path of radius $a < r < b$:	$\displaystyle\oint_C \vec{B}\cdot d\vec{\ell} = B2\pi r = \mu_0 I_C$ or, because $I_C = NI$, $B2\pi r = \mu_0 NI \Rightarrow B = \dfrac{\mu_0 NI}{2\pi r}$

Express the flux in a strip of height H and width dr :	$d\phi_m = BHdr$

Substituting for B yields:	$d\phi_m = \dfrac{\mu_0 NIH}{2\pi r}\,dr$

Integrate $d\phi_m$ from $r = a$ to $r = b$ to obtain:	$\phi_m = \dfrac{\mu_0 NIH}{2\pi}\displaystyle\int_a^b \dfrac{dr}{r} = \dfrac{\mu_0 NIH}{2\pi}\ln\left(\dfrac{b}{a}\right)$

Substitute for ϕ_m in equation (1) and simplify to obtain:	$L = \boxed{\dfrac{\mu_0 N^2 H}{2\pi}\ln\left(\dfrac{b}{a}\right)}$

Magnetic Energy

55 •   In a plane electromagnetic wave, the magnitudes of the electric fields and magnetic fields are related by $E = cB$, where $c = 1/\sqrt{\epsilon_0\,\mu_0}$ is the speed of light. Show that when $E = cB$ the electric and the magnetic energy densities are equal.

Picture the Problem We can examine the ratio of u_m to u_E with $E = cB$ and $c = 1/\sqrt{\epsilon_0\,\mu_0}$ to show that the electric and magnetic energy densities are equal.

Express the ratio of the energy density in the magnetic field to the energy density in the electric field:

$$\frac{u_m}{u_E} = \frac{\dfrac{B^2}{2\mu_0}}{\tfrac{1}{2}\epsilon_0\,E^2} = \frac{B^2}{\mu_0\,\epsilon_0\,E^2}$$

Because $E = cB$:

$$\frac{u_m}{u_E} = \frac{B^2}{\mu_0\,\epsilon_0\,c^2 B^2} = \frac{1}{\mu_0\,\epsilon_0\,c^2}$$

Substituting for c^2 and simplifying yields:

$$\frac{u_m}{u_E} = \frac{\mu_0\,\epsilon_0}{\mu_0\,\epsilon_0} = 1 \Rightarrow u_m = \boxed{u_E}$$

RL Circuits

59 •   A circuit consists of a coil that has a resistance equal to 8.00 Ω and a self-inductance equal to 4.00 mH, an open switch and an ideal 100-V battery—all connected in series. At $t = 0$ the switch is closed. Find the current and its rate of change at times (*a*) $t = 0$, (*b*) $t = 0.100$ ms, (*c*) $t = 0.500$ ms, and (*d*) $t = 1.00$ ms.

Picture the Problem We can find the current using $I = I_f\left(1 - e^{-t/\tau}\right)$ where $I_f = \mathcal{E}_0/R$ and $\tau = L/R$ and its rate of change by differentiating this expression with respect to time.

Express the dependence of the current on I_f and τ:

$$I = I_f\left(1 - e^{-t/\tau}\right)$$

Evaluating I_f and τ yields:

$$I_f = \frac{\mathcal{E}_0}{R} = \frac{100\,\text{V}}{8.00\,\Omega} = 12.5\,\text{A}$$

and

$$\tau = \frac{L}{R} = \frac{4.00\,\text{mH}}{8.00\,\Omega} = 0.500\,\text{ms}$$

Substitute for I_f and τ to obtain:

$$I = (12.5\,\text{A})\left(1 - e^{-t/0.500\,\text{ms}}\right)$$

Express dI/dt:

$$\frac{dI}{dt} = (12.5\,\text{A})\left(-e^{-t/0.500\,\text{ms}}\right)\left(-2000\,\text{s}^{-1}\right)$$
$$= (25.0\,\text{kA/s})e^{-t/0.500\,\text{ms}}$$

(a) Evaluate I and dI/dt at $t = 0$:

$$I(0) = (12.5\,\text{A})\left(1 - e^0\right) = \boxed{0}$$

and

$$\left.\frac{dI}{dt}\right|_{t=0} = (25.0\,\text{kA/s})e^0 = \boxed{25.0\,\text{kA/s}}$$

(b) Evaluating I and dI/dt at $t = 0.100$ ms yields:

$$I(0.100\,\text{ms}) = (12.5\,\text{A})\left(1 - e^{-0.100\,\text{ms}/0.500\,\text{ms}}\right)$$
$$= \boxed{2.27\,\text{A}}$$

and

$$\left.\frac{dI}{dt}\right|_{t=0.500\,\text{ms}} = (25.0\,\text{kA/s})e^{-0.100\,\text{ms}/0.500\,\text{ms}}$$
$$= \boxed{20.5\,\text{kA/s}}$$

(c) Evaluate I and dI/dt at $t = 0.500$ ms to obtain:

$$I(0.500\,\text{ms}) = (12.5\,\text{A})\left(1 - e^{-0.500\,\text{ms}/0.500\,\text{ms}}\right)$$
$$= \boxed{7.90\,\text{A}}$$

and

$$\left.\frac{dI}{dt}\right|_{t=0.500\,\text{ms}} = (25.0\,\text{kA/s})e^{-0.500\,\text{ms}/0.500\,\text{ms}}$$
$$= \boxed{9.20\,\text{kA/s}}$$

(d) Evaluating I and dI/dt at $t = 1.00$ ms yields:

$$I(1.00\,\text{ms}) = (12.5\,\text{A})\left(1 - e^{-1.00\,\text{ms}/0.500\,\text{ms}}\right)$$
$$= \boxed{10.8\,\text{A}}$$

and

$$\left.\frac{dI}{dt}\right|_{t=1.00\,\text{ms}} = (25.0\,\text{kA/s})e^{-1.00\,\text{ms}/0.500\,\text{ms}}$$
$$= \boxed{3.38\,\text{kA/s}}$$

61 •• In the circuit shown in Figure 28-54, let $\mathcal{E}_0 = 12.0$ V, $R = 3.00\ \Omega$, and $L = 0.600$ H. The switch, which was initially open, is closed at time $t = 0$. At time $t = 0.500$ s, find (a) the rate at which the battery supplies energy, (b) the rate of Joule heating in the resistor, and (c) the rate at which energy is being stored in the inductor.

Picture the Problem We can find the current using $I = I_f\left(1 - e^{-t/\tau}\right)$, where $I_f = \mathcal{E}_0/R$, and $\tau = L/R$, and its rate of change by differentiating this expression with respect to time.

Express the dependence of the current on I_f and τ:

$$I(t) = I_f\left(1 - e^{-t/\tau}\right)$$

Evaluating I_f and τ yields:

$$I_f = \frac{\mathcal{E}_0}{R} = \frac{12.0\,\text{V}}{3.00\,\Omega} = 4.00\,\text{A}$$

and

$$\tau = \frac{L}{R} = \frac{0.600\,\text{H}}{3.00\,\Omega} = 0.200\,\text{s}$$

Substitute for I_f and τ to obtain:

$$I(t) = (4.00\,\text{A})\left(1 - e^{-t/0.200\,\text{s}}\right)$$

Express dI/dt:

$$\frac{dI}{dt} = (4.00\,\text{A})\left(-e^{-t/0.200\,\text{s}}\right)\left(-5.00\,\text{s}^{-1}\right)$$
$$= (20.0\,\text{A/s})e^{-t/0.200\,\text{s}}$$

(a) The rate at which the battery supplies energy is given by:

$$P = I\mathcal{E}_0$$

Substituting for I and $\mathcal{E}_0$ yields:

$$P(t) = (4.00\,\text{A})\left(1 - e^{-t/0.200\,\text{s}}\right)(12.0\,\text{V})$$
$$= (48.0\,\text{W})\left(1 - e^{-t/0.200\,\text{s}}\right)$$

The rate at which the battery supplies energy at $t = 0.500$ s is:

$$P(0.500\,\text{s}) = (48.0\,\text{W})\left(1 - e^{-0.500/0.200\,\text{s}}\right)$$
$$= \boxed{44.1\,\text{W}}$$

(b) The rate of Joule heating is:

$$P_J = I^2 R$$

Substitute for I and R and simplify to obtain:

$$P_J = \left[(4.00\,\text{A})\left(1 - e^{-t/0.200\,\text{s}}\right)\right]^2 (3.00\,\Omega)$$
$$= (48.0\,\text{W})\left(1 - e^{-t/0.200\,\text{s}}\right)^2$$

The rate of Joule heating at $t = 0.500$ s is:

$$P_J(0.500\,\text{s}) = (48.0\,\text{W})\left(1 - e^{-0.500\,\text{s}/0.200\,\text{s}}\right)^2$$
$$= \boxed{40.4\,\text{W}}$$

(c) Use the expression for the magnetic energy stored in an inductor to express the rate at which energy is being stored:

$$\frac{dU_L}{dt} = \frac{d}{dt}\left[\tfrac{1}{2}LI^2\right] = LI\frac{dI}{dt}$$

Substitute for L, I, and dI/dt to obtain:

$$\frac{dU_L}{dt} = (0.600\,\text{H})(4.00\,\text{A})\left(1 - e^{-t/0.200\,\text{s}}\right)(20.0\,\text{A/s})e^{-t/0.200\,\text{s}}$$

$$= (48.0\,\text{W})\left(1 - e^{-t/0.200\,\text{s}}\right)e^{-t/0.200\,\text{s}}$$

Evaluate this expression for t = 0.500 s:

$$\left.\frac{dU_L}{dt}\right|_{t=0.500\,\text{s}} = (48.0\,\text{W})\left(1 - e^{-0.500\,\text{s}/0.200\,\text{s}}\right)e^{-0.500\,\text{s}/0.200\,\text{s}} = \boxed{3.62\,\text{W}}$$

Remarks: Note that, to a good approximation, $dU_L/dt = P - P_J$.

63 ••   A circuit consists of a 4.00-mH coil, a 150-Ω resistor, a 12.0-V ideal battery and an open switch—all connected in series. After the switch is closed: (*a*) What is the initial rate of increase of the current? (*b*) What is the rate of increase of the current when the current is equal to half its steady-state value? (*c*) What is the steady-state value of the current? (*d*) How long does it take for the current to reach 99 percent of its steady state value?

Picture the Problem If the current is initially zero in an *LR* circuit, its value at some later time *t* is given by $I = I_f\left(1 - e^{-t/\tau}\right)$, where $I_f = \mathcal{E}_0/R$ and $\tau = L/R$ is the time constant for the circuit. We can find the rate of increase of the current by differentiating *I* with respect to time and the time for the current to reach any given fraction of its initial value by solving for *t*.

(*a*) Express the current in the circuit as a function of time:

$$I = \frac{\mathcal{E}_0}{R}\left(1 - e^{-t/\tau}\right)$$

Express the initial rate of increase of the current by differentiating this expression with respect to time:

$$\frac{dI}{dt} = \frac{\mathcal{E}_0}{R}\frac{d}{dt}\left(1 - e^{-t/\tau}\right)$$

$$= \frac{\mathcal{E}_0}{R}\left(-e^{-t/\tau}\right)\left(-\frac{1}{\tau}\right) = \frac{\mathcal{E}_0}{\tau R}e^{-\frac{R}{L}t}$$

$$= \frac{\mathcal{E}_0}{L}e^{-\frac{R}{L}t}$$

Evaluate dI/dt at $t = 0$ to obtain:

$$\left.\frac{dI}{dt}\right|_{t=0} = \frac{\mathcal{E}_0}{L}e^0 = \frac{12.0\,\text{V}}{4.00\,\text{mH}} = \boxed{3.00\,\text{kA/s}}$$

(b) When $I = 0.5I_f$:

$$0.5 = 1 - e^{-t/\tau} \Rightarrow e^{-t/\tau} = 0.5$$

Evaluate dI/dt with $e^{-t/\tau} = 0.5$ to obtain:

$$\left.\frac{dI}{dt}\right|_{e^{-t/\tau}=0.5} = 0.5\frac{\mathcal{E}_0}{L} = 0.5\left(\frac{12.0\,\text{V}}{4.00\,\text{mH}}\right)$$

$$= \boxed{1.50\,\text{kA/s}}$$

(c) Calculate I_f from $\mathcal{E}_0$ and R:

$$I_f = \frac{\mathcal{E}_0}{R} = \frac{12.0\,\text{V}}{150\,\Omega} = \boxed{80.0\,\text{mA}}$$

(d) When $I = 0.99I_f$:

$$0.99 = 1 - e^{-t/\tau} \Rightarrow e^{-t/\tau} = 0.01$$

Solving for t and substituting for τ yields:

$$t = -\tau\ln(0.01) = -\frac{L}{R}\ln(0.01)$$

Substitute numerical values and evaluate t:

$$t = -\frac{4.00\,\text{mH}}{150\,\Omega}\ln(0.01) = \boxed{0.123\,\text{ms}}$$

65 ••   Given the circuit shown in Figure 28-55, assume that the inductor has negligible internal resistance and that the switch S has been closed for a long time so that a steady current exists in the inductor. (a) Find the battery current, the current in the 100 Ω resistor, and the current in the inductor. (b) Find the potential drop across the inductor immediately after the switch S is opened. (c) Using a **spreadsheet** program, make graphs of the current in the inductor and the potential drop across the inductor as functions of time for the period during which the switch is open.

Picture the Problem The self-induced emf in the inductor is proportional to the rate at which the current through it is changing. Under steady-state conditions, $dI/dt = 0$ and so the self-induced emf in the inductor is zero. We can use Kirchhoff's loop rule to obtain the current through and the voltage across the inductor as a function of time.

(a) Because, under steady-state conditions, the self-induced emf in the inductor is zero and because the inductor has negligible resistance, we can apply Kirchhoff's loop rule to the loop that includes the source, the 10-Ω resistor, and the 2-H inductor to find the current drawn from the battery and flowing through the inductor and the 10-Ω resistor:

$$10\,\text{V} - (10\,\Omega)I_{10-\Omega} = 0$$

Solving for $I_{10-\Omega}$ yields:

$$I_{10-\Omega} = I_{2-\text{H}} = \boxed{1.0\,\text{A}}$$

By applying Kirchhoff's junction rule at the junction between the resistors, we can conclude that:

$$I_{100-\Omega} = I_{\text{battery}} - I_{2-\text{H}} = \boxed{0}$$

(b) When the switch is opened, the current cannot immediately go to zero in the circuit because of the inductor. For a time, a current will circulate in the circuit loop between the inductor and the 100-Ω resistor. Because the current flowing through this circuit is initially 1 A, the voltage drop across the 100-Ω resistor is initially $\boxed{100\,\text{V.}}$ Conservation of energy (Kirchhoff's loop rule) requires that the voltage drop across the 2-H inductor is $V_{2-\text{H}} = \boxed{100\,\text{V.}}$

(c) Apply Kirchhoff's loop rule to the RL circuit to obtain:

$$L\frac{dI}{dt} + IR = 0$$

The solution to this differential equation is:

$$I(t) = I_0 e^{-\frac{R}{L}t} = I_0 e^{-\frac{t}{\tau}}$$

$$\text{where } \tau = \frac{L}{R} = \frac{2.0\,\text{H}}{100\,\Omega} = 0.020\,\text{s}$$

A spreadsheet program to generate the data for graphs of the current and the voltage across the inductor as functions of time is shown below. The formulas used to calculate the quantities in the columns are as follows:

Cell	Formula/Content	Algebraic Form
B1	2.0	L
B2	100	R
B3	1	I_0
A6	0	t_0
B6	\$B\$3*EXP((−\$B\$2/\$B\$1)*A6)	$I_0 e^{-\frac{R}{L}t}$

	A	B	C
1	$L=$	2	H
2	$R=$	100	ohms
3	$I_0=$	1	A
4			
5	t	$I(t)$	$V(t)$
6	0.000	1.00E+00	100.00
7	0.005	7.79E−01	77.88
8	0.010	6.07E−01	60.65
9	0.015	4.72E−01	47.24
10	0.020	3.68E−01	36.79
11	0.025	2.87E−01	28.65
12	0.030	2.23E−01	22.31
32	0.130	1.50E−03	0.15
33	0.135	1.17E−03	0.12
34	0.140	9.12E−04	0.09
35	0.145	7.10E−04	0.07
36	0.150	5.53E−04	0.06

The following graph of the current in the inductor as a function of time was plotted using the data in columns A and B of the spreadsheet program.

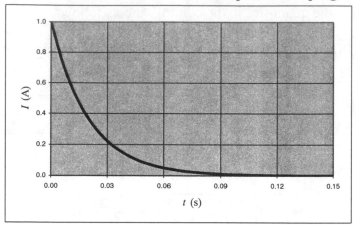

The following graph of the voltage across the inductor as a function of time was plotted using the data in columns A and C of the spreadsheet program.

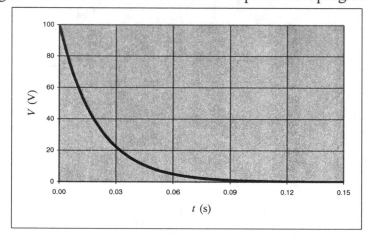

69   •••   In the circuit shown in Figure 28-54, let $\mathcal{E}_0 = 12.0$ V, $R = 3.00\ \Omega$, and $L = 0.600$ H. The switch is closed at time $t = 0$. During the time from $t = 0$ to $t = L/R$, find (*a*) the amount of energy supplied by the battery, (*b*) the amount of energy dissipated in the resistor, and (*c*) the amount of energy delivered to the inductor. *Hint: Find the energy transfer rates as functions of time and integrate.*

Picture the Problem We can integrate $dE/dt = \mathcal{E}_0 I$, where $I = I_{\mathrm{f}}\left(1 - e^{-t/\tau}\right)$, to find the energy supplied by the battery, $dE_{\mathrm{J}}/dt = I^2 R$ to find the energy dissipated in the resistor, and $U_L(\tau) = \tfrac{1}{2}L(I(\tau))^2$ to express the energy that has been stored in the inductor when $t = L/R$.

(a) Express the rate at which energy is supplied by the battery:

$$\frac{dE}{dt} = \mathcal{E}_0 I$$

Express the current in the circuit as a function of time:

$$I = \frac{\mathcal{E}_0}{R}\left(1 - e^{-t/\tau}\right)$$

Substitute for I to obtain:

$$\frac{dE}{dt} = \frac{\mathcal{E}_0^2}{R}\left(1 - e^{-t/\tau}\right)$$

Separate variables and integrate from $t = 0$ to $t = \tau$ to obtain:

$$E = \frac{\mathcal{E}_0^2}{R}\int_0^\tau \left(1 - e^{-t/\tau}\right)dt$$

$$= \frac{\mathcal{E}_0^2}{R}\left[\tau - \left(-\tau e^{-1} + \tau\right)\right]$$

$$= \frac{\mathcal{E}_0^2}{R}\frac{\tau}{e} = \frac{\mathcal{E}_0^2 L}{R^2 e}$$

Substitute numerical values and evaluate E:

$$E = \frac{(12.0\,\text{V})^2(0.600\,\text{H})}{(3.00\,\Omega)^2 e} = \boxed{3.53\,\text{J}}$$

(b) Express the rate at which energy is being dissipated in the resistor:

$$\frac{dE_{\text{J}}}{dt} = I^2 R = \left[\frac{\mathcal{E}_0}{R}\left(1 - e^{-t/\tau}\right)\right]^2 R$$

$$= \frac{\mathcal{E}_0^2}{R}\left(1 - 2e^{-t/\tau} + e^{-2t/\tau}\right)$$

Separate variables and integrate from $t = 0$ to $t = L/R$ to obtain:

$$E_{\text{J}} = \frac{\mathcal{E}_0^2}{R}\int_0^{L/R}\left(1 - 2e^{-t/\tau} + e^{-2t/\tau}\right)dt$$

$$= \frac{\mathcal{E}_0^2}{R}\left(\frac{2\dfrac{L}{R}}{e} - \frac{\dfrac{L}{R}}{2} - \frac{\dfrac{L}{R}}{2e^2}\right)$$

$$= \frac{\mathcal{E}_0^2 L}{R^2}\left(\frac{2}{e} - \frac{1}{2} - \frac{1}{2e^2}\right)$$

Substitute numerical values and evaluate E_{J}:

$$E_{\text{J}} = \frac{(12.0\,\text{V})^2(0.600\,\text{H})}{(3.00\,\Omega)^2}\left(\frac{2}{e} - \frac{1}{2} - \frac{1}{2e^2}\right)$$

$$= \boxed{1.61\,\text{J}}$$

(c) Express the energy stored in the inductor when $t = \dfrac{L}{R}$:

$$U_L\left(\frac{L}{R}\right) = \tfrac{1}{2}L\left(I\left(\frac{L}{R}\right)\right)$$

$$= \tfrac{1}{2}L\left(\frac{\mathcal{E}_0}{R}\left(1-e^{-1}\right)\right)^2$$

$$= \frac{L\mathcal{E}_0^2}{2R^2}\left(1-e^{-1}\right)^2$$

Substitute numerical values and evaluate U_L:

$$U_L\left(\frac{L}{R}\right) = \frac{(0.600\,\mathrm{H})(12.0\,\mathrm{V})^2}{2(3.00\,\Omega)^2}\left(1-e^{-1}\right)^2$$

$$= \boxed{1.92\,\mathrm{J}}$$

Remarks: Note that, as we would expect from energy conservation, $E = E_J + E_L$.

General Problems

71 ••   Figure 28-59 shows a schematic drawing of an *ac generator*. The basic generator consists of a rectangular loop of dimensions a and b and has N turns connected to *slip rings*. The loop rotates (driven by a gasoline engine) at an angular speed of ω in a uniform magnetic field $\vec{B}$. (a) Show that the induced potential difference between the two slip rings is given by $\mathcal{E} = NBab\omega \sin \omega t$. (b) If $a = 2.00$ cm, $b = 4.00$ cm, $N = 250$, and $B = 0.200$ T, at what angular frequency ω must the coil rotate to generate an emf whose maximum value is 100V?

Picture the Problem (*a*) We can apply Faraday's law and the definition of magnetic flux to derive an expression for the induced emf in the coil (potential difference between the slip rings). In Part (*b*) we can solve the equation derived in Part (*a*) for ω and evaluate this expression under the given conditions.

(a) Use Faraday's law to express the induced emf:

$$\mathcal{E} = -\frac{d\phi_m(t)}{dt}$$

Using the definition of magnetic flux, relate the magnetic flux through the loop to its angular velocity:

$$\phi_m(t) = NBA\cos \omega t$$

Substitute for $\phi_m(t)$ to obtain:

$$\varepsilon = -\frac{d}{dt}[NBA\cos\omega t]$$

$$= -NBab\omega(-\sin\omega t)$$

$$= \boxed{NBab\omega\sin\omega t}$$

(b) Express the condition under which $\varepsilon = \varepsilon_{max}$:

$$\sin\omega t = 1$$

and

$$\varepsilon_{max} = NBab\omega \Rightarrow \omega = \frac{\varepsilon_{max}}{NBab}$$

Substitute numerical values and evaluate ω:

$$\omega = \frac{100\,\text{V}}{(250)(0.200\,\text{T})(0.0200\,\text{m})(0.0400\,\text{m})} = \boxed{2.50\,\text{krad/s}}$$

77 •• Figure 28-60*a* shows an experiment designed to measure the acceleration due to gravity. A large plastic tube is encircled by a wire, which is arranged in single loops separated by a distance of 10 cm. A strong magnet is dropped through the top of the loop. As the magnet falls through each loop the voltage rises and then the voltage rapidly falls through zero to a large negative value and then returns to zero. The shape of the voltage signal is shown in Figure 28-62*b*. (*a*) Explain the basic physics behind the generation of this voltage pulse. (*b*) Explain why the tube cannot be made of a conductive material. (*c*) Qualitatively explain the *shape* of the voltage signal in Figure 28-60*b*. (*d*) The times at which the voltage crosses zero as the magnet falls through each loop in succession are given in the table in the next column. Use these data to calculate a value for *g*.

Picture the Problem
(*a*) As the magnet passes through a loop it induces an emf because of the changing flux through the loop. This allows the coil to "sense" when the magnet is passing through it.

(*b*) One cannot use a cylinder made of conductive material because eddy currents induced in it by a falling magnet would slow the magnet.

(*c*) As the magnet approaches the loop the flux increases, resulting in the negative voltage signal of increasing magnitude. When the magnet is passing a loop, the flux reaches a maximum value and then decreases, so the induced emf becomes zero and then positive. The instant at which the induced emf is zero is the instant at which the magnet is at the center of the loop.

(*d*) Each time represents a point when the distance has increased by 10 cm. The following graph of distance versus time was plotted using a spreadsheet program. The regression curve, obtained using Excel's "Add Trendline" feature, is shown as a dashed line.

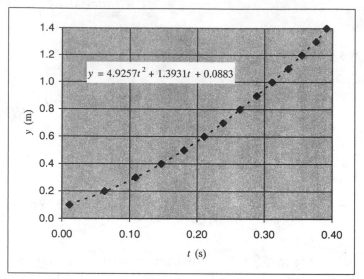

The coefficient of the second-degree term is $\frac{1}{2}g$. Consequently,

$$g = 2(4.9257\,\text{m/s}^2) = \boxed{9.85\,\text{m/s}^2}$$

79 ••   A long solenoid has *n* turns per unit length and carries a current that varies with time according to $I = I_0 \sin \omega t$. The solenoid has a circular cross section of radius *R*. Find the induced electric field, at points near the plane equidistant from the ends of the solenoid, as a function of both the time *t* and the perpendicular distance *r* from the axis of the solenoid for (*a*) *r* < *R* and (*b*) *r* > *R*.

Picture the Problem We can apply Faraday's law to relate the induced electric field *E* to the rates at which the magnetic flux is changing at distances *r* < *R* and *r* > *R* from the axis of the solenoid.

(*a*) Apply Faraday's law to relate the induced electric field to the magnetic flux in the solenoid within a cylindrical region of radius *r* < *R*:

$$\oint_C \vec{E} \cdot d\vec{\ell} = -\frac{d\phi_m}{dt}$$

or

$$E(2\pi r) = -\frac{d\phi_m}{dt} \qquad (1)$$

Express the field within the solenoid:

$$B = \mu_0 n I$$

Express the magnetic flux through an area for which *r* < *R*:

$$\phi_m = BA = \pi r^2 \mu_0 n I$$

Substitute in equation (1) to obtain:

$$E(2\pi r) = -\frac{d}{dt}\left[\pi r^2 \mu_0 nI\right]$$

$$= -\pi r^2 \mu_0 n \frac{dI}{dt}$$

Because $I = I_0 \sin \omega t$:

$$E_{r<R} = -\tfrac{1}{2} r \mu_0 n \frac{d}{dt}\left[I_0 \sin \omega t\right]$$

$$= \boxed{-\tfrac{1}{2} r \mu_0 n I_0 \omega \cos \omega t}$$

(b) Proceed as in (a) with $r > R$ to obtain:

$$E(2\pi r) = -\frac{d}{dt}\left[\pi R^2 \mu_0 nI\right]$$

$$= -\pi R^2 \mu_0 n \frac{dI}{dt}$$

$$= -\pi R^2 \mu_0 n I_0 \omega \cos \omega t$$

Solving for $E_{r>R}$ yields:

$$E_{r>R} = \boxed{-\frac{\mu_0 n R^2 I_0 \omega}{2r} \cos \omega t}$$

Chapter 29
Alternating-Current Circuits

Conceptual Problems

3 • If the frequency in the circuit shown in Figure 29-27 is doubled, the inductance of the inductor will (*a*) double, (*b*) not change, (*c*) halve, (*d*) quadruple.

Determine the Concept The inductance of an inductor is determined by the details of its construction and is independent of the frequency of the circuit. The inductive reactance, on the other hand, is frequency dependent. $\boxed{(b)}$ is correct.

7 • (*a*) In a circuit consisting of a generator and a capacitor, are there any time intervals when the capacitor receives energy from the generator? If so, when? Explain your answer. (*b*) Are there any time intervals when the capacitor supplies power to the generator? If so, when? Explain your answer.

Determine the Concept Yes to both questions. (*a*) While the magnitude of the charge is accumulating on either plate of the capacitor, the capacitor absorbs power from the generator. (*b*) When the magnitude of the charge is on either plate of the capacitor is decreasing, it supplies power to the generator.

9 • Suppose you increase the rotation rate of the coil in the generator shown in the simple ac circuit in Figure 29-29. Then the rms current (*a*) increases, (*b*) does not change, (*c*) may increase or decrease depending on the magnitude of the original frequency, (*d*) may increase or decrease depending on the magnitude of the resistance, (*e*) decreases.

Determine the Concept Because the rms current through the resistor is given by

$$I_{rms} = \frac{\mathcal{E}_{rms}}{R} = \frac{\mathcal{E}_{peak}}{\sqrt{2}} = \frac{NBA}{\sqrt{2}}\omega \, , \; I_{rms} \text{ is directly proportional to } \omega. \; \boxed{(a)} \text{ is correct.}$$

11 • Consider a circuit consisting solely of an ideal inductor and an ideal capacitor. How does the maximum energy stored in the capacitor compare to the maximum value stored in the inductor? (*a*) They are the same and each equal to the total energy stored in the circuit. (*b*) They are the same and each equal to half of the total energy stored in the circuit. (*c*) The maximum energy stored in the capacitor is larger than the maximum energy stored in the inductor. (*d*) The maximum energy stored in the inductor is larger than the maximum energy stored in the capacitor. (*e*) You cannot compare the maximum energies based on the data given because the ratio of the maximum energies depends on the actual capacitance and inductance values.

Determine the Concept The maximum energy stored in the electric field of the capacitor is given by $U_e = \dfrac{1}{2}\dfrac{Q^2}{C}$ and the maximum energy stored in the magnetic field of the inductor is given by $U_m = \dfrac{1}{2}LI^2$. Because energy is conserved in an LC circuit and oscillates between the inductor and the capacitor, $U_e = U_m = U_{total}$. $\boxed{(a)}$ is correct.

17 • True or false:

(*a*) A transformer is used to change frequency.
(*b*) A transformer is used to change voltage.
(*c*) If a transformer steps up the current, it must step down the voltage.
(*d*) A step-up transformer, steps down the current.
(*e*) The standard household wall-outlet voltage in Europe is 220 V, about twice that used in the United States. If a European traveler wants her hair dryer to work properly in the United States, she should use a transformer that has more windings in its secondary coil than in its primary coil.
(*f*) The standard household wall-outlet voltage in Europe is 220 V, about twice that used in the United States. If an American traveler wants his electric razor to work properly in Europe, he should use a transformer that steps up the current.

(*a*) False. A transformer is a device used to raise or lower the voltage in a circuit.

(*b*) True. A transformer is a device used to raise or lower the voltage in a circuit.

(*c*) True. If energy is to be conserved, the product of the current and voltage must be constant.

(*d*) True. Because the product of current and voltage in the primary and secondary circuits is the same, increasing the current in the secondary results in a lowering (or stepping down) of the voltage.

(*e*) True. Because electrical energy is provided at a higher voltage in Europe, the visitor would want to step-up the voltage in order to make her hair dryer work properly.

(*f*) True. Because electrical energy is provided at a higher voltage in Europe, the visitor would want to step-up the current (and decrease the voltage) in order to make his razor work properly.

Alternating Current in Resistors, Inductors, and Capacitors

19 • A 100-W light bulb is screwed into a standard 120-V-rms socket. Find (a) the rms current, (b) the peak current, and (c) the peak power.

Picture the Problem We can use $P_{av} = \varepsilon_{rms} I_{rms}$ to find I_{rms}, $I_{peak} = \sqrt{2} I_{rms}$ to find I_{peak}, and $P_{peak} = I_{peak} \varepsilon_{peak}$ to find P_{peak}.

(a) Relate the average power delivered by the source to the rms voltage across the bulb and the rms current through it:

$$P_{av} = \varepsilon_{rms} I_{rms} \Rightarrow I_{rms} = \frac{P_{av}}{\varepsilon_{rms}}$$

Substitute numerical values and evaluate I_{rms}:

$$I_{rms} = \frac{100\,\text{W}}{120\,\text{V}} = 0.8333\,\text{A} = \boxed{0.833\,\text{A}}$$

(b) Express I_{peak} in terms of I_{rms}:

$$I_{peak} = \sqrt{2} I_{rms}$$

Substitute for I_{rms} and evaluate I_{peak}:

$$I_{peak} = \sqrt{2}(0.8333\,\text{A}) = 1.1785\,\text{A}$$
$$= \boxed{1.18\,\text{A}}$$

(c) Express the maximum power in terms of the maximum voltage and maximum current:

$$P_{peak} = I_{peak} \varepsilon_{peak}$$

Substitute numerical values and evaluate P_{peak}:

$$P_{peak} = (1.1785\,\text{A})\sqrt{2}(120\,\text{V}) = \boxed{200\,\text{W}}$$

21 • What is the reactance of a 1.00-μH inductor at (a) 60 Hz, (b) 600 Hz, and (c) 6.00 kHz?

Picture the Problem We can use $X_L = \omega L$ to find the reactance of the inductor at any frequency.

Express the inductive reactance as a function of f:

$$X_L = \omega L = 2\pi f L$$

(a) At f = 60 Hz:

$$X_L = 2\pi(60\,\text{s}^{-1})(1.00\,\text{mH}) = \boxed{0.38\,\Omega}$$

(b) At $f = 600$ Hz:

$$X_L = 2\pi(600\,\text{s}^{-1})(1.00\,\text{mH}) = \boxed{3.77\,\Omega}$$

(c) At $f = 6.00$ kHz:

$$X_L = 2\pi(6.00\,\text{kHz})(1.00\,\text{mH}) = \boxed{37.7\,\Omega}$$

25 •   A 20-Hz ac generator that produces a peak emf of 10 V is connected to a 20-μF capacitor. Find (a) the peak current and (b) the rms current.

Picture the Problem We can use $I_{peak} = \mathcal{E}_{peak}/X_C$ and $X_C = 1/\omega C$ to express I_{peak} as a function of $\mathcal{E}_{peak}$, f, and C. Once we've evaluate I_{peak}, we can use $I_{rms} = I_{peak}/\sqrt{2}$ to find I_{rms}.

Express I_{peak} in terms of $\mathcal{E}_{peak}$ and X_C:

$$I_{peak} = \frac{\mathcal{E}_{peak}}{X_C}$$

Express the capacitive reactance:

$$X_C = \frac{1}{\omega C} = \frac{1}{2\pi f C}$$

Substitute for X_C and simplify to obtain:

$$I_{peak} = 2\pi f C \mathcal{E}_{peak}$$

(a) Substitute numerical values and evaluate I_{peak}:

$$I_{peak} = 2\pi(20\,\text{s}^{-1})(20\,\mu\text{F})(10\,\text{V})$$
$$= 25.1\,\text{mA} = \boxed{25\,\text{mA}}$$

(b) Express I_{rms} in terms of I_{peak}:

$$I_{rms} = \frac{I_{peak}}{\sqrt{2}} = \frac{25.1\,\text{mA}}{\sqrt{2}} = \boxed{18\,\text{mA}}$$

Undriven Circuits Containing Capacitors, Resistors and Inductors

29 •   (a) What is the period of oscillation of an LC circuit consisting of an ideal 2.0-mH inductor and a 20-μF capacitor? (b) A circuit that oscillates consists solely of an 80-μF capacitor and a variable ideal inductor. What inductance is needed in order to tune this circuit to oscillate at 60 Hz?

Picture the Problem We can use $T = 2\pi/\omega$ and $\omega = 1/\sqrt{LC}$ to relate T (and hence f) to L and C.

(a) Express the period of oscillation of the LC circuit:

$$T = \frac{2\pi}{\omega}$$

For an LC circuit:

$$\omega = \frac{1}{\sqrt{LC}}$$

Substitute for ω to obtain:

$$T = 2\pi\sqrt{LC} \qquad (1)$$

Substitute numerical values and evaluate T:

$$T = 2\pi\sqrt{(2.0\,\text{mH})(20\,\mu\text{F})} = \boxed{1.3\,\text{ms}}$$

(b) Solve equation (1) for L to obtain:

$$L = \frac{T^2}{4\pi^2 C} = \frac{1}{4\pi^2 f^2 C}$$

Substitute numerical values and evaluate L:

$$L = \frac{1}{4\pi^2 \left(60\,\text{s}^{-1}\right)^2 (80\,\mu\text{F})} = \boxed{88\,\text{mH}}$$

33 ••• An inductor and a capacitor are connected, as shown in Figure 29-30. Initially, the switch is open, the left plate of the capacitor has charge Q_0. The switch is then closed. (a) Plot both Q versus t and I versus t on the same graph, and explain how it can be seen from these two plots that the current leads the charge by 90°. (b) The expressions for the charge and for the current are given by Equations 29-38 and 29-39, respectively. Use trigonometry and algebra to show that the current leads the charge by 90°.

Picture the Problem Let Q represent the instantaneous charge on the capacitor and apply Kirchhoff's loop rule to obtain the differential equation for the circuit. We can then solve this equation to obtain an expression for the charge on the capacitor as a function of time and, by differentiating this expression with respect to time, an expression for the current as a function of time. We'll use a spreadsheet program to plot the graphs.

Apply Kirchhoff's loop rule to a clockwise loop just after the switch is closed:

$$\frac{Q}{C} + L\frac{dI}{dt} = 0$$

Because $I = dQ/dt$:

$$L\frac{d^2Q}{dt^2} + \frac{Q}{C} = 0 \text{ or } \frac{d^2Q}{dt^2} + \frac{1}{LC}Q = 0$$

The solution to this equation is:

$$Q(t) = Q_0 \cos(\omega t - \delta)$$

$$\text{where } \omega = \sqrt{\frac{1}{LC}}$$

Because $Q(0) = Q_0$, $\delta = 0$ and: $\qquad\qquad Q(t) = Q_0 \cos \omega t$

The current in the circuit is the derivative of Q with respect to t: $\qquad I = \dfrac{dQ}{dt} = \dfrac{d}{dt}\left[Q_0 \cos \omega t\right] = -\omega Q_0 \sin \omega t$

(*a*) A spreadsheet program was used to plot the following graph showing both the charge on the capacitor and the current in the circuit as functions of time. *L*, *C*, and Q_0 were all arbitrarily set equal to one to obtain these graphs. Note that the current leads the charge by one-fourth of a cycle or 90°.

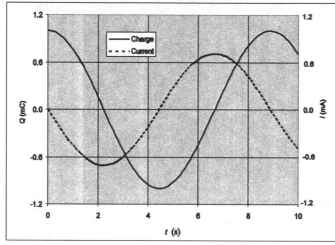

(*b*) The equation for the current is: $\qquad I = -\omega Q_0 \sin \omega t \qquad\qquad$ (1)

The sine and cosine functions are related through the identity: $\qquad -\sin \theta = \cos\left(\theta + \dfrac{\pi}{2}\right)$

Use this identity to rewrite equation (1): $\qquad I = -\omega Q_0 \sin \omega t = \boxed{\omega Q_0 \cos\left(\omega t + \dfrac{\pi}{2}\right)}$

Thus, the current leads the charge by 90°.

Driven *RL* Circuits

35 •• A coil that has a resistance of 80.0 Ω has an impedance of 200 Ω when driven at a frequency of 1.00 kHz. What is the inductance of the coil?

Picture the Problem We can solve the expression for the impedance in an *LR* circuit for the inductive reactance and then use the definition of X_L to find *L*.

Express the impedance of the coil in terms of its resistance and inductive reactance:

$$Z = \sqrt{R^2 + X_L^2}$$

Solve for X_L to obtain:

$$X_L = \sqrt{Z^2 - R^2}$$

Express X_L in terms of L:

$$X_L = 2\pi f L$$

Equate these two expressions to obtain:

$$2\pi f L = \sqrt{Z^2 - R^2} \Rightarrow L = \frac{\sqrt{Z^2 - R^2}}{2\pi f}$$

Substitute numerical values and evaluate L:

$$L = \frac{\sqrt{(200\,\Omega)^2 - (80.0\,\Omega)^2}}{2\pi(1.00\,\text{kHz})}$$

$$= \boxed{29.2\,\text{mH}}$$

39 •• A coil that has a resistance R and an inductance L has a power factor equal to 0.866 when driven at a frequency of 60 Hz. What is the coil's power factor it is driven at 240 Hz?

Picture the Problem We can use the definition of the power factor to find the relationship between X_L and R when the coil is driven at a frequency of 60 Hz and then use the definition of X_L to relate the inductive reactance at 240 Hz to the inductive reactance at 60 Hz. We can then use the definition of the power factor to determine its value at 240 Hz.

Using the definition of the power factor, relate R and X_L:

$$\cos\delta = \frac{R}{Z} = \frac{R}{\sqrt{R^2 + X_L^2}} \qquad (1)$$

Square both sides of the equation to obtain:

$$\cos^2\delta = \frac{R^2}{R^2 + X_L^2}$$

Solve for $X_L^2(60\,\text{Hz})$:

$$X_L^2(60\,\text{Hz}) = R^2\left(\frac{1}{\cos^2\delta} - 1\right)$$

Substitute for $\cos\delta$ and simplify to obtain:

$$X_L^2(60\,\text{Hz}) = R^2\left(\frac{1}{(0.866)^2} - 1\right) = \tfrac{1}{3}R^2$$

Use the definition of X_L to obtain:

$$X_L^2(f) = 4\pi f^2 L^2 \text{ and } X_L^2(f') = 4\pi f'^2 L^2$$

Dividing the second of these equations by the first and simplifying yields:

$$\frac{X_L^2(f')}{X_L^2(f)} = \frac{4\pi f'^2 L^2}{4\pi f^2 L^2} = \frac{f'^2}{f^2}$$

or

$$X_L^2(f') = \left(\frac{f'}{f}\right)^2 X_L^2(f)$$

Substitute numerical values to obtain:

$$X_L^2(240\,\text{Hz}) = \left(\frac{240\,\text{s}^{-1}}{60\,\text{s}^{-1}}\right)^2 X_L^2(60\,\text{Hz})$$

$$= 16\left(\frac{1}{3}R^2\right) = \frac{16}{3}R^2$$

Substitute in equation (1) to obtain:

$$(\cos\delta)_{240\,\text{Hz}} = \frac{R}{\sqrt{R^2 + \frac{16}{3}R^2}} = \sqrt{\frac{3}{19}}$$

$$= \boxed{0.397}$$

41 ••   Figure 29-33 shows a load resistor that has a resistance of $R_L = 20.0\ \Omega$ connected to a high-pass filter consisting of an inductor that has inductance $L = 3.20$-mH and a resistor that has resistance $R = 4.00$-Ω. The output of the ideal ac generator is given by $\mathcal{E} = (100\ \text{V})\cos(2\pi ft)$. Find the rms currents in all three branches of the circuit if the driving frequency is (*a*) 500 Hz and (*b*) 2000 Hz. Find the fraction of the total average power supplied by the ac generator that is delivered to the load resistor if the frequency is (*c*) 500 Hz and (*d*) 2000 Hz.

Picture the Problem $\mathcal{E} = V_1 + V_2$, where V_1 is the voltage drop across R and V_2 is the voltage drop across the parallel combination of L and R_L. $\vec{\mathcal{E}} = \vec{V}_1 + \vec{V}_2$ is the relation for the phasors. For the parallel combination $\vec{I} = \vec{I}_{R_L} + \vec{I}_L$. Also, V_1 is in phase with I and V_2 is in phase with I_{R_L}. First draw the phasor diagram for the currents in the parallel combination, then add the phasors for the voltages to the diagram.

The phasor diagram for the currents in the circuit is:

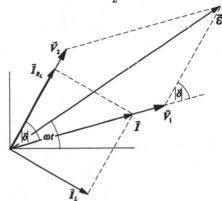

Adding the voltage phasors to the diagram gives:

The maximum current in the inductor, $I_{2,\,peak}$, is given by:

$$I_{2,\,peak} = \frac{V_{2,\,peak}}{Z_2} \qquad (1)$$

where $Z_2^{-2} = R_L^{-2} + X_L^{-2}$ $\qquad (2)$

$\tan|\delta|$ is given by:

$$\tan|\delta| = \frac{I_{L,\,peak}}{I_{R,\,peak}} = \frac{V_{2,\,peak}/X_L}{V_{2,\,peak}/R_L}$$

$$= \frac{R_L}{X_L} = \frac{R_L}{\omega L} = \frac{R_L}{2\pi f L}$$

Solve for $|\delta|$ to obtain:

$$|\delta| = \tan^{-1}\left(\frac{R_L}{2\pi f L}\right) \qquad (3)$$

Apply the law of cosines to the triangle formed by the voltage phasors to obtain:

$$\mathcal{E}_{peak}^2 = V_{1,\,peak}^2 + V_{2,\,peak}^2 + 2V_{1,\,peak}V_{2,\,peak}\cos|\delta|$$

or

$$I_{peak}^2 Z^2 = I_{peak}^2 R^2 + I_{peak}^2 Z_2^2 + 2I_{peak}RI_{peak}Z_2\cos|\delta|$$

Dividing out the current squared yields:

$$Z^2 = R^2 + Z_2^2 + 2RZ_2\cos|\delta|$$

Solving for Z yields:

$$Z = \sqrt{R^2 + Z_2^2 + 2RZ_2 \cos|\delta|} \quad (4)$$

The maximum current I_{peak} in the circuit is given by:

$$I_{peak} = \frac{\mathcal{E}_{peak}}{Z} \quad (5)$$

I_{rms} is related to I_{peak} according to:

$$I_{rms} = \frac{1}{\sqrt{2}} I_{peak} \quad (6)$$

(*a*) Substitute numerical values in equation (3) and evaluate $|\delta|$:

$$|\delta| = \tan^{-1}\left(\frac{20.0\,\Omega}{2\pi(500\,\text{Hz})(3.20\,\text{mH})}\right)$$

$$= \tan^{-1}\left(\frac{20.0\,\Omega}{10.053\,\Omega}\right) = 63.31°$$

Solving equation (2) for Z_2 yields:

$$Z_2 = \frac{1}{\sqrt{R_L^{-2} + X_L^{-2}}}$$

Substitute numerical values and evaluate Z_2:

$$Z_2 = \frac{1}{\sqrt{(20.0\,\Omega)^{-2} + (10.053\,\Omega)^{-2}}}$$

$$= 8.982\,\Omega$$

Substitute numerical values and evaluate Z:

$$Z = \sqrt{(4.00\,\Omega)^2 + (8.982\,\Omega)^2 + 2(4.00\,\Omega)(8.982\,\Omega)\cos 63.31°} = 11.36\,\Omega$$

Substitute numerical values in equation (5) and evaluate I_{peak}:

$$I_{peak} = \frac{100\,\text{V}}{11.36\,\Omega} = 8.806\,\text{A}$$

Substitute for I_{peak} in equation (6) and evaluate I_{rms}:

$$I_{rms} = \frac{1}{\sqrt{2}}(8.806\,\text{A}) = \boxed{6.23\,\text{A}}$$

The maximum and rms values of V_2 are given by:

$$V_{2,peak} = I_{peak}Z_2$$

$$= (8.806\,\text{A})(8.982\,\Omega) = 79.095\,\text{V}$$

and

$$V_{2,rms} = \frac{1}{\sqrt{2}}V_{2,peak}$$

$$= \frac{1}{\sqrt{2}}(79.095\,\text{V}) = 55.929\,\text{V}$$

The rms values of $I_{R_L,\text{rms}}$ and $I_{L,\text{rms}}$ are:

$$I_{R_L,\text{rms}} = \frac{V_{2,\text{rms}}}{R_L} = \frac{55.929\,\text{V}}{20.0\,\Omega} = \boxed{2.80\,\text{A}}$$

and

$$I_{L,\text{rms}} = \frac{V_{2,\text{rms}}}{X_L} = \frac{55.929\,\text{V}}{10.053\,\Omega} = \boxed{5.53\,\text{A}}$$

(b) Proceed as in (a) with $f = 2000$ Hz to obtain:

$X_L = 40.2\,\Omega, |\delta| = 26.4°, Z_2 = 17.9\,\Omega,$

$Z = 21.6\,\Omega, I_{\text{peak}} = 4.64\,\text{A},$ and

$I_{\text{rms}} = \boxed{3.28\,\text{A}},$

$V_{2,\text{max}} = 83.0\,\text{V}, V_{2,\text{rms}} = 58.7\,\text{V},$

$I_{R_L,\text{rms}} = \boxed{2.94\,\text{A}},$ and $I_{L,\text{rms}} = \boxed{1.46\,\text{A}}$

(c) The power delivered by the ac source equals the sum of the power dissipated in the two resistors. The fraction of the total power delivered by the source that is dissipated in load resistor is given by:

$$\frac{P_{R_L}}{P_{R_L} + P_R} = \left(1 + \frac{P_R}{P_{R_L}}\right)^{-1} = \left(1 + \frac{I_{\text{rms}}^2 R}{I_{R_L,\text{rms}}^2 R_L}\right)^{-1}$$

Substitute numerical values for $f = 500$ Hz to obtain:

$$\frac{P_{R_L}}{P_{R_L} + P_R}\bigg|_{f=500\,\text{Hz}} = \left(1 + \frac{(6.23\,\text{A})^2 (4.00\,\Omega)}{(2.80\,\text{A})^2 (20.0\,\Omega)}\right)^{-1} = 0.502 = \boxed{50.2\%}$$

(d) Substitute numerical values for $f = 2000$ Hz to obtain:

$$\frac{P_{R_L}}{P_{R_L} + P_R}\bigg|_{f=2000\,\text{Hz}} = \left(1 + \frac{(3.28\,\text{A})^2 (4.00\,\Omega)}{(2.94\,\text{A})^2 (20.0\,\Omega)}\right)^{-1} = 0.800 = \boxed{80.0\%}$$

Filters and Rectifiers

47   ••   A slowly varying voltage signal $V(t)$ is applied to the input of the high-pass filter of Problem 44. Slowly varying means that during one time constant (equal to RC) there is no significant change in the voltage signal. Show that under these conditions the output voltage is proportional to the time derivative of $V(t)$. This situation is known as a *differentiation circuit*.

Picture the Problem We can use Kirchhoff's loop rule to obtain a differential equation relating the input, capacitor, and resistor voltages. Because the voltage drop across the resistor is small compared to the voltage drop across the capacitor, we can express the voltage drop across the capacitor in terms of the input voltage.

Apply Kirchhoff's loop rule to the input side of the filter to obtain:

$$V(t) - V_C - IR = 0$$

where V_C is the potential difference across the capacitor.

Substitute for $V(t)$ and I to obtain:

$$V_{\text{in peak}} \cos \omega t - V_c - R\frac{dQ}{dt} = 0$$

Because $Q = CV_C$:

$$\frac{dQ}{dt} = \frac{d}{dt}[CV_C] = C\frac{dV_C}{dt}$$

Substitute for dQ/dt to obtain:

$$V_{\text{peak}} \cos \omega t - V_C - RC\frac{dV_C}{dt} = 0$$

the differential equation describing the potential difference across the capacitor.

Because there is no significant change in the voltage signal during one time constant:

$$\frac{dV_C}{dt} = 0 \Rightarrow RC\frac{dV_C}{dt} = 0$$

Substituting for $RC\dfrac{dV_C}{dt}$ yields:

$$V_{\text{in peak}} \cos \omega t - V_C = 0$$
and
$$V_C = V_{\text{in peak}} \cos \omega t$$

Consequently, the potential difference across the resistor is given by:

$$V_R = RC\frac{dV_C}{dt} = \boxed{RC\frac{d}{dt}\left[V_{\text{in peak}} \cos \omega t\right]}$$

49 •• Show that the average power dissipated in the resistor of the high-pass filter of Problem 44 is given by $P_{\text{ave}} = \dfrac{V_{\text{in peak}}^2}{2R\left[1 + (\omega RC)^{-2}\right]}$.

Picture the Problem We can express the instantaneous power dissipated in the resistor and then use the fact that the average value of the square of the cosine function over one cycle is ½ to establish the given result.

The instantaneous power $P(t)$ dissipated in the resistor is:

$$P(t) = \frac{V_{\text{out}}^2}{R}$$

The output voltage V_{out} is:

$$V_{out} = V_H \cos(\omega t - \delta)$$

From Problem 44:

$$V_H = \frac{V_{in\ peak}}{\sqrt{1 + (\omega RC)^{-2}}}$$

Substitute in the expression for $P(t)$ to obtain:

$$P(t) = \frac{V_H^2}{R} \cos^2(\omega t - \delta)$$

$$= \frac{V_{in\ peak}^2}{R[1 + (\omega RC)^{-2}]} \cos^2(\omega t - \delta)$$

Because the average value of the square of the cosine function over one cycle is ½:

$$\boxed{P_{ave} = \frac{V_{in\ peak}^2}{2R[1 + (\omega RC)^{-2}]}}$$

51 •• The circuit shown in Figure 29-36 is an example of a low-pass filter. (Assume that the output is connected to a load that draws only an insignificant amount of current.) (a) If the input voltage is given by $V_{in} = V_{in\ peak} \cos \omega t$, show that the output voltage is $V_{out} = V_L \cos(\omega t - \delta)$ where $V_L = V_{in\ peak} \big/ \sqrt{1 + (\omega RC)^2}$. (b) Discuss the trend of the output voltage in the limiting cases $\omega \to 0$ and $\omega \to \infty$.

Picture the Problem In the phasor diagram for the RC low-pass filter, $\vec{V}_{app}$ and $\vec{V}_C$ are the phasors for V_{in} and V_{out}, respectively. The projection of $\vec{V}_{app}$ onto the horizontal axis is $V_{app} = V_{in}$, the projection of $\vec{V}_C$ onto the horizontal axis is $V_C = V_{out}$, $V_{peak} = |\vec{V}_{app}|$, and ϕ is the angle between $\vec{V}_C$ and the horizontal axis.

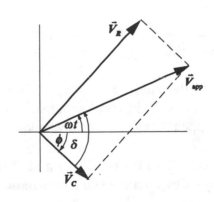

(a) Express V_{app}:

$$V_{app} = V_{in\ peak} \cos \omega t$$
$$\text{where } V_{in\ peak} = I_{peak} Z$$
$$\text{and } Z^2 = R^2 + X_C^2 \qquad (1)$$

$V_{out} = V_C$ is given by:

$$V_{out} = V_{C,peak} \cos \phi$$
$$= I_{peak} X_C \cos \phi$$

If we define δ as shown in the phasor diagram, then:

$$V_{out} = I_{peak} X_C \cos(\omega t - \delta)$$

$$= \frac{V_{in\ peak}}{Z} X_C \cos(\omega t - \delta)$$

Solving equation (1) for Z and substituting for X_C yields:

$$Z = \sqrt{R^2 + \left(\frac{1}{\omega C}\right)^2} \qquad (2)$$

Using equation (2) to substitute for Z and substituting for X_C yields:

$$V_{out} = \frac{V_{in\ peak}}{\sqrt{R^2 + \left(\frac{1}{\omega C}\right)^2}} \frac{1}{\omega C} \cos(\omega t - \delta)$$

Simplify further to obtain:

$$V_{out} = \frac{V_{in\ peak}}{\sqrt{1 + (\omega RC)^2}} \cos(\omega t - \delta)$$

or

$$V_{out} = \boxed{V_L \cos(\omega t - \delta)}$$

where

$$V_L = \boxed{\frac{V_{in\ peak}}{\sqrt{1 + (\omega RC)^2}}}$$

(b) Note that, as $\omega \to 0$, $V_L \to V_{peak}$. This makes sense physically in that, for low frequencies, X_C is large and, therefore, a larger peak input voltage will appear across it than appears across it for high frequencies.

Note further that, as $\omega \to \infty$, $V_L \to 0$. This makes sense physically in that, for high frequencies, X_C is small and, therefore, a smaller peak voltage will appear across it than appears across it for low frequencies.

Remarks: In Figures 29-19 and 29-20, δ is defined as the phase of the voltage drop across the combination relative to the voltage drop across the resistor.

55 ••• The circuit shown in Figure 29-37 is a *trap filter*. (Assume that the output is connected to a load that draws only an insignificant amount of current.) (*a*) Show that the *trap filter* acts to reject signals in a band of frequencies centered at $\omega = 1/\sqrt{LC}$. (*b*) How does the width of the frequency band rejected depend on the resistance R?

Picture the Problem The phasor diagram for the *trap* filter is shown below. $\vec{V}_{app}$ and $\vec{V}_L + \vec{V}_C$ are the phasors for V_{in} and V_{out}, respectively. The projection of $\vec{V}_{app}$ onto the horizontal axis is $V_{app} = V_{in}$, and the projection of $\vec{V}_L + \vec{V}_C$ onto the horizontal axis is $V_L + V_C = V_{out}$. Requiring that the impedance of the trap be zero will yield the frequency at which the circuit rejects signals. Defining the bandwidth as $\Delta\omega = |\omega - \omega_{trap}|$ and requiring that $|Z_{trap}| = R$ will yield an expression for the bandwidth and reveal its dependence on R.

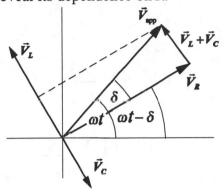

(a) Express V_{app} :

$$V_{app} = V_{app,\,peak} \cos \omega t$$

where $V_{app,\,peak} = V_{peak} = I_{peak}Z$

and $Z^2 = R^2 + (X_L - X_C)^2$ (1)

V_{out} is given by:

$$V_{out} = V_{out,\,peak} \cos(\omega t - \delta)$$

where $V_{out,\,peak} = I_{peak}Z_{trap}$

and $Z_{trap} = X_L - X_C$

Solving equation (1) for Z yields:

$$Z = \sqrt{R^2 + (X_L - X_C)^2}$$ (2)

Because $V_{out} = V_L + V_C$:

$$\begin{aligned}
V_{out} &= V_{out,\,peak} \cos(\omega t - \delta) \\
&= I_{peak}Z_{trap} \cos(\omega t - \delta) \\
&= \frac{V_{peak}}{Z} Z_{trap} \cos(\omega t - \delta)
\end{aligned}$$

Using equation (2) to substitute for Z yields:

$$V_{out} = \frac{V_{peak}}{\sqrt{R^2 + Z_{trap}^2}} Z_{trap} \cos(\omega t - \delta)$$

Noting that $V_{out} = 0$ provided $Z_{trap} = 0$, set $Z_{trap} = 0$ to obtain:

$$Z_{trap} = X_L - X_C = 0$$

Substituting for X_L and X_C yields:

$$\omega L - \frac{1}{\omega C} = 0 \Rightarrow \omega = \boxed{\frac{1}{\sqrt{LC}}}$$

(b) Let the bandwidth $\Delta\omega$ be:

$$\Delta\omega = |\omega - \omega_{trap}| \qquad (3)$$

Let the frequency bandwidth be defined by the frequency at which $|Z_{trap}| = R$. Then:

$$\omega L - \frac{1}{\omega C} = R \Rightarrow \omega^2 LC - 1 = \omega RC$$

Because $\omega_{trap} = \frac{1}{\sqrt{LC}}$:

$$\left(\frac{\omega}{\omega_{trap}}\right)^2 - 1 = \omega RC$$

For $\omega \approx \omega_{trap}$:

$$\left(\frac{\omega^2 - \omega_{trap}^2}{\omega_{trap}}\right) \approx \omega_{trap} RC$$

Solve for $\omega^2 - \omega_{trap}^2$:

$$\omega^2 - \omega_{trap}^2 = (\omega - \omega_{trap})(\omega + \omega_{trap})$$

Because $\omega \approx \omega_{trap}$, $\omega + \omega_{trap} \approx 2\omega_{trap}$:

$$\omega^2 - \omega_{trap}^2 \approx 2\omega_{trap}(\omega - \omega_{trap})$$

Substitute in equation (3) to obtain:

$$\Delta\omega = |\omega - \omega_{trap}| = \frac{RC\omega_{trap}^2}{2} = \boxed{\frac{R}{2L}}$$

Driven *RLC* Circuits

63 •• Show that the expression $P_{av} = R\mathcal{E}_{rms}^2/Z^2$ gives the correct result for a circuit containing only an ideal ac generator and (a) a resistor, (b) a capacitor, and (c) an inductor. In the expression $P_{av} = R\mathcal{E}_{rms}^2/Z^2$, P_{av} is the average power supplied by the generator, $\mathcal{E}_{rms}$ is the root-mean-square of the emf of the generator, R is the resistance, C is the capacitance and L is the inductance. (In Part (a), $C = L = 0$, in Part (b), $R = L = 0$ and in Part (c), $R = C = 0$.

Picture the Problem The impedance of an ac circuit is given by $Z = \sqrt{R^2 + (X_L - X_C)^2}$. We can evaluate the given expression for P_{av} first for $X_L = X_C = 0$ and then for $R = 0$.

(a) For $X = 0$, $Z = R$ and:

$$P_{av} = \frac{R\mathcal{E}_{rms}^2}{Z^2} = \frac{R\mathcal{E}_{rms}^2}{R^2} = \boxed{\frac{\mathcal{E}_{rms}^2}{R}}$$

(b) and (c) If $R = 0$, then:

$$P_{av} = \frac{R\mathcal{E}_{rms}^2}{Z^2} = \frac{(0)\mathcal{E}_{rms}^2}{(X_L - X_C)^2} = \boxed{0}$$

Remarks: Recall that there is no energy dissipation in an ideal inductor or capacitor.

65 ••   Find (*a*) the Q factor and (*b*) the resonance width (in hertz) for the circuit in Problem 64. (*c*) What is the power factor when $\omega = 8000$ rad/s?

Picture the Problem The Q factor of the circuit is given by $Q = \omega_0 L/R$, the resonance width by $\Delta f = f_0/Q = \omega_0/2\pi Q$, and the power factor by $\cos\delta = R/Z$. Because Z is frequency dependent, we'll need to find X_C and X_L at $\omega = 8000$ rad/s in order to evaluate $\cos\delta$.

Using their definitions, express the Q factor and the resonance width of the circuit:

$$Q = \frac{\omega_0 L}{R} \qquad (1)$$

and

$$\Delta f = \frac{f_0}{Q} = \frac{\omega_0}{2\pi Q} \qquad (2)$$

(*a*) Express the resonance frequency for the circuit:

$$\omega_0 = \frac{1}{\sqrt{LC}}$$

Substituting for ω_0 in equation (1) yields:

$$Q = \frac{L}{\sqrt{LC}\,R} = \frac{1}{R}\sqrt{\frac{L}{C}}$$

Substitute numerical values and evaluate Q:

$$Q = \frac{1}{5.0\,\Omega}\sqrt{\frac{10\,\text{mH}}{2.0\,\mu\text{F}}} = 14.1 = \boxed{14}$$

(*b*) Substitute numerical values in equation (2) and evaluate Δf:

$$\Delta f = \frac{7.07\times10^3\,\text{rad/s}}{2\pi(14.1)} = \boxed{80\,\text{Hz}}$$

(*c*) The power factor of the circuit is given by:

$$\cos\delta = \frac{R}{Z} = \frac{R}{\sqrt{R^2 + (X_L - X_C)^2}} = \frac{R}{\sqrt{R^2 + \left(\omega L - \dfrac{1}{\omega C}\right)^2}}$$

Substitute numerical values and evaluate $\cos\delta$:

$$\cos\delta = \frac{5.0\,\Omega}{\sqrt{(5.0\,\Omega)^2 + \left((8000\,\text{s}^{-1})(10\,\text{mH}) - \dfrac{1}{(8000\,\text{s}^{-1})(2.0\,\mu\text{F})}\right)^2}} = \boxed{0.27}$$

69 ••   In the circuit shown in Figure 29-42 the ideal generator produces an rms voltage of 115 V when operated at 60 Hz. What is the rms voltage between points (*a*) *A* and *B*, (*b*) *B* and *C*, (*c*) *C* and *D*, (*d*) *A* and *C*, and (*e*) *B* and *D*?

Picture the Problem We can find the rms current in the circuit and then use it to find the potential differences across each of the circuit elements. We can use phasor diagrams and our knowledge of the phase shifts between the voltages across the three circuit elements to find the voltage differences across their combinations.

(*a*) Express the potential difference between points *A* and *B* in terms of I_{rms} and X_L:

$$V_{AB} = I_{rms} X_L \qquad (1)$$

Express I_{rms} in terms of $\mathcal{E}$ and *Z*:

$$I_{rms} = \frac{\mathcal{E}}{Z} = \frac{\mathcal{E}}{\sqrt{R^2 + (X_L - X_C)^2}}$$

Evaluate X_L and X_C to obtain:

$$X_L = 2\pi f L = 2\pi(60\,\text{s}^{-1})(137\,\text{mH})$$
$$= 51.648\,\Omega$$

and

$$X_C = \frac{1}{2\pi f C} = \frac{1}{2\pi(60\,\text{s}^{-1})(25\,\mu\text{F})}$$
$$= 106.10\,\Omega$$

Substitute numerical values and evaluate I_{rms}:

$$I_{rms} = \frac{115\,\text{V}}{\sqrt{(50\,\Omega)^2 + (51.648\,\Omega - 106.10\,\Omega)^2}}$$
$$= 1.5556\,\text{A}$$

Substitute numerical values in equation (1) and evaluate V_{AB}:

$$V_{AB} = (1.5556\,\text{A})(51.648\,\Omega) = 80.344\,\text{V}$$
$$= \boxed{80\,\text{V}}$$

(*b*) Express the potential difference between points *B* and *C* in terms of I_{rms} and *R*:

$$V_{BC} = I_{rms} R = (1.5556\,\text{A})(50\,\Omega)$$
$$= 77.780\,\text{V} = \boxed{78\,\text{V}}$$

(*c*) Express the potential difference between points *C* and *D* in terms of I_{rms} and X_C:

$$V_{CD} = I_{rms} X_C = (1.5556\,\text{A})(106.10\,\Omega)$$
$$= 165.05\,\text{V} = \boxed{0.17\,\text{kV}}$$

(d) The voltage across the inductor leads the voltage across the resistor as shown in the phasor diagram to the right:

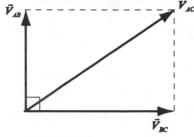

Use the Pythagorean theorem to find V_{AC}:

$$V_{AC} = \sqrt{V_{AB}^2 + V_{BC}^2}$$
$$= \sqrt{(80.0\,\text{V})^2 + (77.780\,\text{V})^2}$$
$$= 111.58\,\text{V} = \boxed{0.11\,\text{kV}}$$

(e) The voltage across the capacitor lags the voltage across the resistor as shown in the phasor diagram to the right:

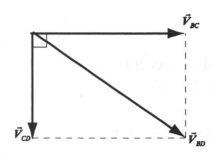

Use the Pythagorean theorem to find V_{BD}:

$$V_{BD} = \sqrt{V_{CD}^2 + V_{BC}^2}$$
$$= \sqrt{(165.05\,\text{V})^2 + (77.780\,\text{V})^2}$$
$$= 182.46\,\text{V} = \boxed{0.18\,\text{kV}}$$

The Transformer

79 • A rms voltage of 24 V is required for a device whose impedance is 12 Ω. (a) What should the turns ratio of a transformer be, so that the device can be operated from a 120-V line? (b) Suppose the transformer is accidentally connected in reverse with the secondary winding across the 120-V-rms line and the 12-Ω load across the primary. How much rms current will then be in the primary winding?

Picture the Problem Let the subscript 1 denote the primary and the subscript 2 the secondary. We can use $V_2 N_1 = V_1 N_2$ and $N_1 I_1 = N_2 I_2$ to find the turns ratio and the primary current when the transformer connections are reversed.

(a) Relate the number of primary and secondary turns to the primary and secondary voltages:

$$V_{2,\,\text{rms}} N_1 = V_{1,\,\text{rms}} N_2 \qquad (1)$$

Solve for and evaluate the ratio N_2/N_1:

$$\frac{N_2}{N_1} = \frac{V_{2,\,rms}}{V_{1,\,rms}} = \frac{24\,V}{120\,V} = \boxed{\frac{1}{5}}$$

(b) Relate the current in the primary to the current in the secondary and to the turns ratio:

$$I_{1,\,rms} = \frac{N_2}{N_1} I_{2,\,rms}$$

Express the current in the primary winding in terms of the voltage across it and its impedance:

$$I_{2,\,rms} = \frac{V_{2,\,rms}}{Z_2}$$

Substitute for $I_{2,\,rms}$ to obtain:

$$I_{1,\,rms} = \frac{N_2}{N_1} \frac{V_{2,\,rms}}{Z_2}$$

Substitute numerical values and evaluate $I_{1,\,rms}$:

$$I_1 = \left(\frac{5}{1}\right)\left(\frac{120\,V}{12\,\Omega}\right) = \boxed{50\,A}$$

General Problems

85 ••   Figure 29-45 shows the voltage versus time for a *square-wave* voltage source. If $V_0 = 12$ V, (a) what is the rms voltage of this source?
(b) If this alternating waveform is rectified by eliminating the negative voltages, so that only the positive voltages remain, what is the new rms voltage?

Picture the Problem The average of any quantity over a time interval ΔT is the integral of the quantity over the interval divided by ΔT. We can use this definition to find both the average of the voltage squared, $\left(V^2\right)_{av}$ and then use the definition of the rms voltage.

(a) From the definition of V_{rms} we have:

$$V_{rms} = \sqrt{\left(V_0^2\right)_{av}}$$

Noting that $-V_0^2 = V_0^2$, evaluate V_{rms}:

$$V_{rms} = \sqrt{V_0^2} = V_0 = \boxed{12\,V}$$

(b) Noting that the voltage during the second half of each cycle is now zero, express the voltage during the first half cycle of the time interval $\frac{1}{2}\Delta T$:

$$V = V_0$$

Express the square of the voltage during this half cycle:

$$V^2 = V_0^2$$

Calculate $\left(V^2\right)_{av}$ by integrating V^2 from $t = 0$ to $t = \frac{1}{2}\Delta T$ and dividing by ΔT:

$$\left(V^2\right)_{av} = \frac{V_0^2}{\Delta T}\int_0^{\frac{1}{2}\Delta T}dt = \frac{V_0^2}{\Delta T}[t]_0^{\frac{1}{2}\Delta T} = \frac{1}{2}V_0^2$$

Substitute to obtain:

$$V_{rms} = \sqrt{\frac{1}{2}V_0^2} = \frac{V_0}{\sqrt{2}} = \frac{12\,V}{\sqrt{2}} = \boxed{8.5\,V}$$

86 •• What are the average values and rms values of current for the two current waveforms shown in Figure 29-46?

Picture the Problem The average of any quantity over a time interval ΔT is the integral of the quantity over the interval divided by ΔT. We can use this definition to find both the average current I_{av}, and the average of the current squared, $\left(I^2\right)_{av}$

From the definition of I_{av} and I_{rms} we have:

$$I_{av} = \frac{1}{\Delta T}\int_0^{\Delta T} I\,dt \text{ and } I_{rms} = \sqrt{\left(I^2\right)_{av}}$$

Waveform (a) Express the current during the first half cycle of time interval ΔT:

$$I_a = \frac{4\,A}{\Delta T}t$$

where I is in A when t and T are in seconds.

Evaluate $I_{av,\,a}$:

$$I_{av,\,a} = \frac{1}{\Delta T}\int_0^{\Delta T}\frac{4.0\,A}{\Delta T}t\,dt = \frac{4.0\,A}{\left(\Delta T\right)^2}\int_0^{\Delta T}t\,dt$$

$$= \frac{4.0\,A}{\left(\Delta T\right)^2}\left[\frac{t^2}{2}\right]_0^{\Delta T} = \boxed{2.0\,A}$$

Express the square of the current during this half cycle:

$$I_a^2 = \frac{\left(4.0\,A\right)^2}{\left(\Delta T\right)^2}t^2$$

Noting that the average value of the squared current is the same for each time interval ΔT, calculate $\left(I_a^2\right)_{av}$ by integrating I_a^2 from $t = 0$ to $t = \Delta T$ and dividing by ΔT:

$$\left(I_a^2\right)_{av} = \frac{1}{\Delta T}\int_0^{\Delta T}\frac{(4.0\,\text{A})^2}{(\Delta T)^2}t^2\,dt$$

$$= \frac{(4.0\,\text{A})^2}{(\Delta T)^3}\left[\frac{t^3}{3}\right]_0^{\Delta T} = \frac{16}{3}\,\text{A}^2$$

Substitute in the expression for $I_{rms,a}$ to obtain:

$$I_{rms,a} = \sqrt{\frac{16}{3}\,\text{A}^2} = \boxed{2.3\,\text{A}}$$

Waveform (b) Noting that the current during the second half of each cycle is zero, express the current during the first half cycle of the time interval $\frac{1}{2}\Delta T$:

$$I_b = 4.0\,\text{A}$$

Evaluate $I_{av,b}$:

$$I_{av,b} = \frac{4.0\,\text{A}}{\Delta T}\int_0^{\frac{1}{2}\Delta T}dt = \frac{4.0\,\text{A}}{\Delta T}\left[t\right]_0^{\frac{1}{2}\Delta T}$$

$$= \boxed{2.0\,\text{A}}$$

Express the square of the current during this half cycle:

$$I_b^2 = (4.0\,\text{A})^2$$

Calculate $\left(I_b^2\right)_{av}$ by integrating I_b^2 from $t = 0$ to $t = \frac{1}{2}\Delta T$ and dividing by ΔT:

$$\left(I_b^2\right)_{av} = \frac{(4.0\,\text{A})^2}{\Delta T}\int_0^{\frac{1}{2}\Delta}dt$$

$$= \frac{(4.0\,\text{A})^2}{\Delta T}\left[t\right]_0^{\frac{1}{2}\Delta T} = 8.0\,\text{A}^2$$

Substitute in the expression for $I_{rms,b}$ to obtain:

$$I_{rms,b} = \sqrt{8.0\,\text{A}^2} = \boxed{2.8\,\text{A}}$$

87 ••   In the circuit shown in Figure 29-47, $\mathcal{E}_1 = (20\,\text{V})\cos 2\pi ft$, where $f = 180$ Hz; $\mathcal{E}_2 = 18$ V, and $R = 36\,\Omega$. Find the maximum, minimum, average, and rms values of the current in the resistor.

Picture the Problem We can apply Kirchhoff's loop rule to express the current in the circuit in terms of the emfs of the sources and the resistance of the resistor. We can then find I_{max} and I_{min} by considering the conditions under which the time-dependent factor in I will be a maximum or a minimum. Finally, we can use

$I_{rms} = \sqrt{\left(I^2\right)_{av}}$ to derive an expression for I_{rms} that we can use to determine its value.

Apply Kirchhoff's loop rule to obtain:

$$\mathcal{E}_{1,\,peak} \cos \omega t + \mathcal{E}_2 - IR = 0$$

Solving for I yields:

$$I = \frac{\mathcal{E}_{1,\,peak}}{R} \cos \omega t + \frac{\mathcal{E}_2}{R}$$

or

$$I = A_1 \cos \omega t + A_2$$

where $A_1 = \dfrac{\mathcal{E}_{1,\,peak}}{R}$ and $A_2 = \dfrac{\mathcal{E}_2}{R}$

Substitute numerical values to obtain:

$$I = \left(\frac{20\,V}{36\,\Omega}\right)\cos\left(2\pi\left(180\,s^{-1}\right)t\right) + \frac{18\,V}{36\,\Omega}$$

$$= (0.556\,A)\cos\left(1131\,s^{-1}\right)t + 0.50\,A$$

The current is a maximum when $\cos\left(1131\,s^{-1}\right)t = 1$. Hence :

$$I_{max} = 0.50\,A + 0.556\,A = \boxed{1.06\,A}$$

Evaluate I_{min} :

$$I_{min} = 0.50\,A - 0.556\,A = \boxed{-0.06\,A}$$

Because the average value of $\cos \omega t = 0$:

$$I_{av} = \boxed{0.50\,A}$$

The rms current is the square root of the average of the squared current:

$$I_{rms} = \sqrt{\left[I^2\right]_{av}} \qquad (1)$$

$\left[I^2\right]_{av}$ is given by:

$$\left[I^2\right]_{av} = \left[\left(A_1 \cos \omega t + A_2\right)^2\right]_{av}$$

$$= \left[A_1^2 \cos^2 \omega t + 2A_1 A_2 \cos \omega t + A_2^2\right]_{av}$$

$$= \left[A_1^2 \cos^2 \omega t\right]_{av} + \left[2A_1 A_2 \cos \omega t\right]_{av} + \left[A_2^2\right]_{av}$$

$$= A_1^2 \left[\cos^2 \omega t\right]_{av} + 2A_1 A_2 \left[\cos \omega t\right]_{av} + A_2^2$$

Because $\left[\cos^2 \omega t\right]_{av} = \tfrac{1}{2}$ and $\left[\cos \omega t\right]_{av} = 0$:

$$\left[I^2\right]_{av} = \tfrac{1}{2}A_1^2 + A_2^2$$

Substituting in equation (1) yields:

$$I_{rms} = \sqrt{\tfrac{1}{2}A_1^2 + A_2^2}$$

Substitute for A_1 and A_2 to obtain:

$$I_{rms} = \sqrt{\frac{1}{2}\left(\frac{\mathcal{E}_1}{R}\right)^2 + \left(\frac{\mathcal{E}_2}{R}\right)^2}$$

Substitute numerical values and evaluate I_{rms}:

$$I_{rms} = \sqrt{\frac{1}{2}\left(\frac{20\text{ V}}{36\,\Omega}\right)^2 + \left(\frac{18\text{ V}}{36\,\Omega}\right)^2}$$

$$= \boxed{0.64\text{ A}}$$

89 ••• A circuit consists of an ac generator, a capacitor and an ideal inductor—all connected in series. The emf of the generator is given by $\mathcal{E}_{peak} \cos\omega t$. (a) Show that the charge on the capacitor obeys the equation

$$L\frac{d^2Q}{dt^2} + \frac{Q}{C} = \mathcal{E}_{peak} \cos\omega t.$$ (b) Show by direct substitution that this equation is

satisfied by $Q = Q_{peak} \cos\omega t$ where $Q_{peak} = -\dfrac{\mathcal{E}_{peak}}{L(\omega^2 - \omega_0^2)}$. (c) Show that the current

can be written as $I = I_{peak} \cos(\omega t - \delta)$, where $I_{peak} = \dfrac{\omega\mathcal{E}_{peak}}{L|\omega^2 - \omega_0^2|} = \dfrac{\mathcal{E}_{peak}}{|X_L - X_C|}$ and

$\delta = -90°$ for $\omega < \omega_0$ and $\delta = 90°$ for $\omega > \omega_0$, where ω_0 is the resonance frequency.

Picture the Problem In Part (a) we can apply Kirchhoff's loop rule to obtain the 2nd order differential equation relating the charge on the capacitor to the time. In Part (b) we'll assume a solution of the form $Q = Q_{peak} \cos\omega t$, differentiate it twice, and substitute for d^2Q/dt^2 and Q to show that the assumed solution satisfies the differential equation provided $Q_{peak} = -\dfrac{\mathcal{E}_{peak}}{L(\omega^2 - \omega_0^2)}$. In Part (c) we'll use our results from (a) and (b) to establish the result for I_{peak} given in the problem statement.

(a) Apply Kirchhoff's loop rule to obtain:

$$\mathcal{E} - \frac{Q}{C} - L\frac{dI}{dt} = 0$$

Substitute for $\mathcal{E}$ and rearrange the differential equation to obtain:

$$L\frac{dI}{dt} + \frac{Q}{C} = \mathcal{E}_{max} \cos\omega t$$

Because $I = dQ/dt$:

$$\boxed{L\frac{d^2Q}{dt^2} + \frac{Q}{C} = \mathcal{E}_{max} \cos\omega t}$$

(b) Assume that the solution is:

$$Q = Q_{peak} \cos\omega t$$

Differentiate the assumed solution twice to obtain:

$$\frac{dQ}{dt} = -\omega Q_{peak} \sin \omega t$$

and

$$\frac{d^2Q}{dt^2} = -\omega^2 Q_{peak} \cos \omega t$$

Substitute for $\frac{dQ}{dt}$ and $\frac{d^2Q}{dt^2}$ in the differential equation to obtain:

$$-\omega^2 L Q_{peak} \cos \omega t + \frac{Q_{peak}}{C} \cos \omega t$$
$$= \mathcal{E}_{peak} \cos \omega t$$

Factor $\cos \omega t$ from the left-hand side of the equation:

$$\left(-\omega^2 L Q_{peak} + \frac{Q_{peak}}{C} \right) \cos \omega t$$
$$= \mathcal{E}_{peak} \cos \omega t$$

If this equation is to hold for all values of t it must be true that:

$$-\omega^2 L Q_{peak} + \frac{Q_{peak}}{C} = \mathcal{E}_{peak}$$

Solving for Q_{peak} yields:

$$Q_{peak} = \frac{\mathcal{E}_{peak}}{-\omega^2 L + \frac{1}{C}}$$

Factor L from the denominator and substitute for $1/LC$ to obtain:

$$Q_{peak} = \frac{\mathcal{E}_{peak}}{L\left(-\omega^2 + \frac{1}{LC} \right)}$$

$$= \boxed{ -\frac{\mathcal{E}_{peak}}{L\left(\omega^2 - \omega_0^2 \right)} }$$

(c) From (a) and (b) we have:

$$I = \frac{dQ}{dt} = -\omega Q_{peak} \sin \omega t$$

$$= \frac{\omega \mathcal{E}_{peak}}{L\left(\omega^2 - \omega_0^2\right)} \sin \omega t = I_{peak} \sin \omega t$$

$$= \boxed{I_{peak} \cos\left(\omega t - \delta\right)}$$

where

$$I_{peak} = \boxed{\frac{\omega \mathcal{E}_{peak}}{L\left|\omega^2 - \omega_0^2\right|}} = \frac{\mathcal{E}_{peak}}{\frac{L}{\omega}\left|\omega^2 - \omega_0^2\right|}$$

$$= \frac{\mathcal{E}_{peak}}{\left|\omega L - \dfrac{1}{\omega C}\right|} = \boxed{\frac{\mathcal{E}_{peak}}{\left|X_L - X_C\right|}}$$

If $\omega > \omega_0$, $X_L > X_C$ and the current lags the voltage by 90° ($\delta = 90°$).

If $\omega < \omega_0$, $X_L < X_C$ and the current leads the voltage by 90° ($\delta = -90°$).

Chapter 30
Maxwell's Equations and Electromagnetic Waves

Conceptual Problems

1 •   True or false:

(a) The displacement current has different units than the conduction current.
(b) Displacement current only exists if the electric field in the region is changing with time.
(c) In an oscillating LC circuit, no displacement current exists between the capacitor plates when the capacitor is momentarily fully charged.
(d) In an oscillating LC circuit, no displacement current exists between the capacitor plates when the capacitor is momentarily uncharged.

(a) False. Like those of conduction current, the units of displacement current are C/s.

(b) True. Because displacement current is given by $I_d = \epsilon_0 \, d\phi_e/dt$, I_d is zero if $d\phi_e/dt = 0$.

(c) True. When the capacitor is fully charged, the electric flux is momentarily a maximum (its rate of change is zero) and, consequently, the displacement current between the plates of the capacitor is zero.

(d) False. I_d is zero if $d\phi_e/dt = 0$. At the moment when the capacitor is momentarily uncharged, $dE/dt \neq 0$ and so $d\phi_e/dt \neq 0$.

3 •   True or false:

(a)   Maxwell's equations apply only to electric and magnetic fields that are constant over time.
(b)   The electromagnetic wave equation can be derived from Maxwell's equations.
(c)   Electromagnetic waves are transverse waves.
(d)   The electric and magnetic fields of an electromagnetic wave in free space are in phase.

(a) False. Maxwell's equations apply to both time-independent and time-dependent fields.

(b) True. One can use Faraday's law and the modified version of Ampere's law to derive the wave equation.

(c) True. Both the electric and magnetic fields of an electromagnetic wave oscillate at right angles to the direction of propagation of the wave.

(d) True.

9 • If a red light beam, a green light beam, and a violet light beam, all traveling in empty space, have the same intensity, which light beam carries more momentum? (a) the red light beam, (b) the green light beam, (c) the violet light beam, (d) They all have the same momentum. (e) You cannot determine which beam carries the most momentum from the data given.

Determine the Concept The momentum of an electromagnetic wave is directly proportional to its energy ($p = U/c$). Because the intensity of a wave is its energy per unit area and per unit time (the average value of its Poynting vector), waves with equal intensity have equal energy and equal momentum. $\boxed{(d)}$ is correct.

Estimation and Approximation

13 •• One of the first successful satellites launched by the United States in the 1950s was essentially a large spherical (aluminized) Mylar balloon from which radio signals were reflected. After several orbits around Earth, scientists noticed that the orbit itself was changing with time. They eventually determined that radiation pressure from the sunlight was causing the orbit of this object to change—a phenomenon not taken into account in planning the mission. Estimate the ratio of the radiation-pressure force by the sunlight on the satellite to the gravitational force by Earth's gravity on the satellite.

Picture the Problem We can use the definition of pressure to express the radiation force on the balloon. We'll assume that the gravitational force on the balloon is approximately its weight at the surface of the Earth, that the density of Mylar is approximately that of water and that the area receiving the radiation from the sunlight is the cross-sectional area of the balloon.

The radiation force acting on the balloon is given by:

$$F_r = P_r A$$

where A is the cross-sectional area of the balloon.

Because the radiation from the Sun is reflected, the radiation pressure is twice what it would be if it were absorbed:

$$P_r = \frac{2I}{c}$$

Substituting for P_r and A yields:

$$F_r = \frac{2I\left(\frac{1}{4}\pi d^2\right)}{c} = \frac{\pi d^2 I}{2c}$$

The gravitational force acting on the balloon when it is in a near-Earth orbit is approximately its weight at the surface of Earth:

$$F_g = w_{balloon} = m_{balloon}g = \rho_{Mylar}V_{Mylar}g$$
$$= \rho_{Mylar}A_{surface, ballon}t\,g$$

where t is the thickness of the Mylar skin of the balloon.

Because the surface area of the balloon is $4\pi r^2 = \pi d^2$:

$$F_g = \pi\rho_{Mylar}d^2 t\,g$$

Express the ratio of the radiation-pressure force to the gravitational force and simplify to obtain:

$$\frac{F_r}{F_g} = \frac{\dfrac{\pi d^2 I}{2c}}{\pi\rho_{Mylar}d^2 t\,g} = \frac{I}{2\rho_{Mylar}t\,gc}$$

Assuming the thickness of the Mylar skin of the balloon to be 1 mm, substitute numerical values and evaluate F_r/F_g:

$$\frac{F_r}{F_g} = \frac{1.35\dfrac{kW}{m^2}}{2\left(1.00\times10^3\dfrac{kg}{m^3}\right)\left(9.81\dfrac{m}{s^2}\right)(1\,mm)\left(2.998\times10^8\dfrac{m}{s}\right)} \approx \boxed{2\times10^{-7}}$$

Maxwell's Displacement Current

15 •   A parallel-plate capacitor has circular plates and no dielectric between the plates. Each plate has a radius equal to 2.3 cm and the plates are separated by 1.1 mm. Charge is flowing onto the upper plate (and off of the lower plate) at a rate of 5.0 A. (*a*) Find the rate of change of the electric field strength in the region between the plates. (*b*) Compute the displacement current in the region between the plates and show that it equals 5.0 A.

Picture the Problem We can differentiate the expression for the electric field between the plates of a parallel-plate capacitor to find the rate of change of the electric field strength and the definitions of the conduction current and electric flux to compute I_d.

(*a*) Express the electric field strength between the plates of the parallel-plate capacitor:

$$E = \frac{Q}{\epsilon_0 A}$$

Differentiate this expression with respect to time to obtain an expression for the rate of change of the electric field strength:

$$\frac{dE}{dt} = \frac{d}{dt}\left[\frac{Q}{\epsilon_0 A}\right] = \frac{1}{\epsilon_0 A}\frac{dQ}{dt} = \frac{I}{\epsilon_0 A}$$

Substitute numerical values and evaluate dE/dt:

$$\frac{dE}{dt} = \frac{5.0\,\text{A}}{\left(8.854\times10^{-12}\,\text{C}^2/\text{N}\cdot\text{m}^2\right)\pi(0.023\,\text{m})^2} = 3.40\times10^{14}\,\text{V/m}\cdot\text{s}$$

$$= \boxed{3.4\times10^{14}\,\text{V/m}\cdot\text{s}}$$

(b) Express the displacement current I_d:

$$I_d = \epsilon_0\frac{d\phi_e}{dt}$$

Substitute for the electric flux to obtain:

$$I_d = \epsilon_0\frac{d}{dt}[EA] = \epsilon_0 A\frac{dE}{dt}$$

Substitute numerical values and evaluate I_d:

$$I_d = \left(8.854\times10^{-12}\,\text{C}^2/\text{N}\cdot\text{m}^2\right)\pi(0.023\,\text{m})^2(3.40\times10^{14}\,\text{V/m}\cdot\text{s}) = \boxed{5.0\,\text{A}}$$

19 ••   There is a current of 10 A in a resistor that is connected in series with a parallel plate capacitor. The plates of the capacitor have an area of 0.50 m², and no dielectric exists between the plates. (a) What is the displacement current between the plates? (b) What is the rate of change of the electric field strength between the plates? (c) Find the value of the line integral $\oint_C \vec{B}\cdot d\vec{\ell}$, where the integration path C is a 10-cm-radius circle that lies in a plane that is parallel with the plates and is completely within the region between them.

Picture the Problem We can use the conservation of charge to find I_d, the definitions of the displacement current and electric flux to find dE/dt, and Ampere's law to evaluate $\vec{B}\cdot d\vec{\ell}$ around the given path.

(a) From conservation of charge we know that:

$$I_d = I = \boxed{10\,\text{A}}$$

(b) Express the displacement current I_d:

$$I_d = \epsilon_0\frac{d\phi_e}{dt} = \epsilon_0\frac{d}{dt}[EA] = \epsilon_0 A\frac{dE}{dt}$$

Substituting for dE/dt yields:

$$\frac{dE}{dt} = \frac{I_d}{\epsilon_0 A}$$

Substitute numerical values and evaluate dE/dt:

$$\frac{dE}{dt} = \frac{10\,\text{A}}{\left(8.85 \times 10^{-12} \, \dfrac{\text{C}^2}{\text{N} \cdot \text{m}^2}\right)\left(0.50\,\text{m}^2\right)}$$

$$= \boxed{2.3 \times 10^{12} \, \frac{\text{V}}{\text{m} \cdot \text{s}}}$$

(c) Apply Ampere's law to a circular path of radius r between the plates and parallel to their surfaces to obtain:

$$\oint_C \vec{B} \cdot d\vec{\ell} = \mu_0 I_{\text{enclosed}}$$

Assuming that the displacement current is uniformly distributed and letting A represent the area of the circular plates yields:

$$\frac{I_{\text{enclosed}}}{\pi r^2} = \frac{I_d}{A} \Rightarrow I_{\text{enclosed}} = \frac{\pi r^2}{A} I_d$$

Substitute for I_{enclosed} to obtain:

$$\oint_C \vec{B} \cdot d\vec{\ell} = \frac{\mu_0 \pi r^2}{A} I_d$$

Substitute numerical values and evaluate $\oint_C \vec{B} \cdot d\vec{\ell}$:

$$\oint_C \vec{B} \cdot d\vec{\ell} = \frac{\left(4\pi \times 10^{-7} \, \text{N}/\text{A}^2\right)\pi\left(0.10\,\text{m}\right)^2\left(10\,\text{A}\right)}{0.50\,\text{m}^2} = \boxed{0.79 \, \mu\text{T} \cdot \text{m}}$$

Electric Dipole Radiation

27 ••• A radio station that uses a vertical electric dipole antenna broadcasts at a frequency of 1.20 MHz and has a total power output of 500 kW. Calculate the intensity of the signal at a horizontal distance of 120 km from the station.

Picture the Problem The intensity of radiation from an electric dipole is given by $C(\sin^2\theta)/r^2$, where C is a constant whose units are those of power, r is the distance from the dipole to the point of interest, and θ is the angle between the antenna and the position vector $\vec{r}$. We can integrate the intensity to express the total power radiated by the antenna and use this result to evaluate C. Knowing C we can find the intensity at a horizontal distance of 120 km.

Express the intensity of the signal as a function of r and θ:

$$I(r,\theta)=C\,\frac{\sin^2\theta}{r^2}$$

At a horizontal distance of 120 km from the station:

$$I(120\,\text{km},90°)=C\,\frac{\sin^2 90°}{(120\,\text{km})^2}$$

$$=\frac{C}{(120\,\text{km})^2} \tag{1}$$

From the definition of intensity we have:

$$dP = I\,dA$$

and

$$P_{\text{tot}} = \iint I(r,\theta)\,dA$$

where, in polar coordinates,

$$dA = r^2 \sin\theta\,d\theta\,d\phi$$

Substitute for dA to obtain:

$$P_{\text{tot}} = \int_0^{2\pi}\int_0^{\pi} I(r,\theta)\,r^2 \sin\theta\,d\theta\,d\phi$$

Substitute for $I(r,\theta)$:

$$P_{\text{tot}} = C\int_0^{2\pi}\int_0^{\pi} \sin^3\theta\,d\theta\,d\phi$$

From integral tables we find that:

$$\int_0^{\pi}\sin^3\theta\,d\theta = -\tfrac{1}{3}\cos\theta\big(\sin^2\theta+2\big)\Big]_0^{\pi} = \frac{4}{3}$$

Substitute and integrate with respect to ϕ to obtain:

$$P_{\text{tot}} = \frac{4}{3}C\int_0^{2\pi}d\phi = \frac{4}{3}C[\phi]_0^{2\pi} = \frac{8\pi}{3}C$$

Solving for C yields:

$$C = \frac{3}{8\pi}P_{\text{tot}}$$

Substitute for P_{tot} and evaluate C to obtain:

$$C = \frac{3}{8\pi}(500\,\text{kW}) = 59.68\,\text{kW}$$

Substituting for C in equation (1) and evaluating $I(120\ \text{km},90°)$:

$$I(120\,\text{km},90°)=\frac{59.68\,\text{kW}}{(120\,\text{km})^2}$$

$$= \boxed{4.14\,\mu\text{W/m}^2}$$

Energy and Momentum in an Electromagnetic Wave

31 •   The amplitude of an electromagnetic wave's electric field is
400 V/m. Find the wave's (*a*) rms electric field strength, (*b*) rms magnetic field
strength, (*c*) intensity and (*d*) radiation pressure (P_r).

Picture the Problem The rms values of the electric and magnetic fields are found
from their amplitudes by dividing by the square root of two. The rms values of the
electric and magnetic field strengths are related according to $B_{rms} = E_{rms}/c$. We can
find the intensity of the radiation using $I = E_{rms}B_{rms}/\mu_0$ and the radiation pressure
using $P_r = I/c$.

(*a*) Relate E_{rms} to E_0:

$$E_{rms} = \frac{E_0}{\sqrt{2}} = \frac{400\,\text{V/m}}{\sqrt{2}} = 282.8\,\text{V/m}$$

$$= \boxed{283\,\text{V/m}}$$

(*b*) Find B_{rms} from E_{rms}:

$$B_{rms} = \frac{E_{rms}}{c} = \frac{282.8\,\text{V/m}}{2.998\times10^8\,\text{m/s}}$$

$$= 0.9434\,\mu\text{T} = \boxed{943\ \text{nT}}$$

(*c*) The intensity of an
electromagnetic wave is given by:

$$I = \frac{E_{rms}B_{rms}}{\mu_0}$$

Substitute numerical values and
evaluate *I*:

$$I = \frac{(282.8\,\text{V/m})(0.9434\,\mu\text{T})}{4\pi\times10^{-7}\,\text{N/A}^2}$$

$$= 212.3\,\text{W/m}^2 = \boxed{212\,\text{W/m}^2}$$

(*d*) Express the radiation pressure in
terms of the intensity of the wave:

$$P_r = \frac{I}{c}$$

Substitute numerical values and
evaluate P_r:

$$P_r = \frac{212.3\,\text{W/m}^2}{2.998\times10^8\,\text{m/s}} = \boxed{708\ \text{nPa}}$$

35 •   (*a*) For a given distance from a radiating electric dipole, at
what angle (expressed as θ and measured from the dipole axis) is the intensity
equal to 50 percent of the maximum intensity? (*b*) At what angle θ is the intensity
equal to 1 percent of the maximum intensity?

Picture the Problem At a fixed distance from the electric dipole, the intensity of
radiation is a function θ alone.

(a) The intensity of the radiation from the dipole is proportional to $\sin^2\theta$:

$$I(\theta) = I_0 \sin^2\theta \qquad (1)$$

where I_0 is the maximum intensity.

For $I = \frac{1}{2}I_0$:

$$\tfrac{1}{2}I_0 = I_0\sin^2\theta \Rightarrow \sin^2\theta = \tfrac{1}{2}$$

Solving for θ yields:

$$\theta = \sin^{-1}\left(\sqrt{\tfrac{1}{2}}\right) = \boxed{45^\circ}$$

(b) For $I = 0.01I_0$:

$$0.01I_0 = I_0\sin^2\theta \Rightarrow \sin^2\theta = 0.01$$

Solving for θ yields:

$$\theta = \sin^{-1}\left(\sqrt{0.01}\right) = \boxed{5.7^\circ}$$

37 •• An electromagnetic plane wave has an electric field that is parallel to the y axis, and has a Poynting vector that is given by $\vec{S}(x,t) = \left(100\ \text{W/m}^2\right)\cos^2[kx - \omega t]\,\hat{i}$, where x is in meters, $k = 10.0$ rad/m, $\omega = 3.00 \times 10^9$ rad/s, and t is in seconds. (a) What is the direction of propagation of the wave? (b) Find the wavelength and frequency of the wave. (c) Find the electric and magnetic fields of the wave as functions of x and t.

Picture the Problem We can determine the direction of propagation of the wave, its wavelength, and its frequency by examining the argument of the cosine function. We can find E from $|\vec{S}| = E^2/\mu_0 c$ and B from $B = E/c$. Finally, we can use the definition of the Poynting vector and the given expression for $\vec{S}$ to find $\vec{E}$ and $\vec{B}$.

(a) Because the argument of the cosine function is of the form $kx - \omega t$, the wave propagates in the $+x$ direction.

(b) Examining the argument of the cosine function, we note that the wave number k of the wave is:

$$k = \frac{2\pi}{\lambda} = 10.0\,\text{m}^{-1} \Rightarrow \lambda = \boxed{0.628\,\text{m}}$$

Examining the argument of the cosine function, we note that the angular frequency ω of the wave is:

$$\omega = 2\pi f = 3.00 \times 10^9\,\text{s}^{-1}$$

Solving for f yields:

$$f = \frac{3.00 \times 10^9\,\text{s}^{-1}}{2\pi} = \boxed{477\,\text{MHz}}$$

(c) Express the magnitude of $\vec{S}$ in terms of E:

$$\left|\vec{S}\right| = \frac{E^2}{\mu_0 c} \Rightarrow E = \sqrt{\mu_0 c \left|\vec{S}\right|}$$

Substitute numerical values and evaluate E:

$$E = \sqrt{\left(4\pi \times 10^{-7}\ \text{N/A}^2\right)\left(2.998 \times 10^8\ \text{m/s}\right)\left(100\ \text{W/m}^2\right)} = 194.1\ \text{V/m}$$

Because

$$\vec{S}(x,t) = \left(100\ \text{W/m}^2\right)\cos^2[kx - \omega t]\,\hat{i}$$

and $\vec{S} = \dfrac{1}{\mu_0}\vec{E} \times \vec{B}$:

$$\vec{E}(x,t) = \boxed{\left(194\ \text{V/m}\right)\cos[kx - \omega t]\,\hat{j}}$$

where $k = 10.0$ rad/m and $\omega = 3.00 \times 10^9$ rad/s.

Use $B = E/c$ to evaluate B:

$$B = \frac{194.1\ \text{V/m}}{2.998 \times 10^8\ \text{m/s}} = 647.4\ \text{nT}$$

Because $\vec{S} = \dfrac{1}{\mu_0}\vec{E} \times \vec{B}$, the direction of $\vec{B}$ must be such that the cross product of $\vec{E}$ with $\vec{B}$ is in the positive x direction:

$$\vec{B}(x,t) = \boxed{\left(647\ \text{nT}\right)\cos[kx - \omega t]\,\hat{k}}$$

where $k = 10.0$ rad/m and $\omega = 3.00 \times 10^9$ rad/s.

39 •• A pulsed laser fires a 1000-MW pulse that has a 200-ns duration at a small object that has a mass equal to 10.0 mg and is suspended by a fine fiber that is 4.00 cm long. If the radiation is completely absorbed by the object, what is the maximum angle of deflection of this pendulum? (Think of the system as a ballistic pendulum and assume the small object was hanging vertically before the radiation hit it.)

Picture the Problem The diagram shows the displacement of the pendulum bob, through an angle θ, as a consequence of the complete absorption of the radiation incident on it. We can use conservation of energy (mechanical energy is conserved *after* the collision) to relate the maximum angle of deflection of the pendulum to the initial momentum of the pendulum bob. Because the displacement of the bob during the absorption of the pulse is negligible, we can use conservation of momentum (conserved *during* the collision) to equate the momentum of the electromagnetic pulse to the initial momentum of the bob.

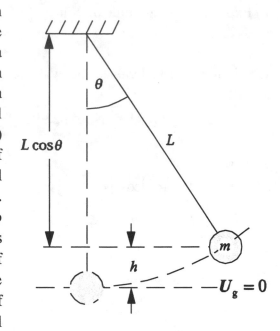

Apply conservation of energy to obtain:

$$K_f - K_i + U_f - U_i = 0$$

or, because $U_i = K_f = 0$ and $K_i = \dfrac{p_i^2}{2m}$,

$$-\frac{p_i^2}{2m} + U_f = 0$$

U_f is given by:

$$U_f = mgh = mgL(1 - \cos\theta)$$

Substitute for U_f:

$$-\frac{p_i^2}{2m} + mgL(1 - \cos\theta) = 0$$

Solve for θ to obtain:

$$\theta = \cos^{-1}\left(1 - \frac{p_i^2}{2m^2 gL}\right)$$

Use conservation of momentum to relate the momentum of the electromagnetic pulse to the initial momentum p_i of the pendulum bob:

$$P_{\text{em wave}} = \frac{U}{c} = \frac{P\Delta t}{c} = p_i$$

where Δt is the duration of the pulse.

Substitute for p_i:

$$\theta = \cos^{-1}\left[1 - \frac{P^2(\Delta t)^2}{2m^2 c^2 gL}\right]$$

Substitute numerical values and evaluate θ:

$$\theta = \cos^{-1}\left[1 - \frac{(1000\,\text{MW})^2(200\,\text{ns})^2}{2(10.0\,\text{mg})^2(2.998\times10^8\,\text{m/s})^2(9.81\,\text{m/s}^2)(0.0400\,\text{m})}\right]$$

$$= \boxed{6.10\times10^{-3}\ \text{degrees}}$$

Remarks: The solution presented here is valid only if the displacement of the bob during the absorption of the pulse is negligible. (Otherwise, the horizontal component of the momentum of the pulse-bob system is not conserved during the collision.) We can show that the displacement during the pulse-bob collision is small by solving for the speed of the bob after absorbing the pulse. Applying conservation of momentum ($mv = P(\Delta t)/c$) and solving for v gives $v = 6.67 \times 10^{-7}$ m/s. This speed is so slow compared to c, we can conclude that the duration of the collision is extremely close to 200 ns (the time for the pulse to travel its own length). Traveling at 6.67×10^{-7} m/s for 200 ns, the bob would travel 1.33×10^{-13} m—a distance 1000 times smaller that the diameter of a hydrogen atom. (Because 6.67×10^{-7} m/s is the maximum speed of the bob during the collision, the bob would actually travel less than 1.33×10^{-13} m during the collision.)

41   ••   (*a*) Estimate the force on Earth due to the pressure of the radiation on Earth by the Sun, and compare this force to the gravitational force of the Sun on Earth. (At Earth's orbit, the intensity of sunlight is 1.37 kW/m².)
(*b*)· Repeat Part (*a*) for Mars which is at an average distance of 2.28×10^8 km from the Sun and has a radius of 3.40×10^3 km. (*c*) Which planet has the larger ratio of radiation pressure to gravitational attraction.

Picture the Problem We can find the radiation pressure force from the definition of pressure and the relationship between the radiation pressure and the intensity of the radiation from the Sun. We can use Newton's law of gravitation to find the gravitational force the Sun exerts on Earth and Mars.

(*a*) The radiation pressure exerted on Earth is given by:

$$P_{\text{r, Earth}} = \frac{F_{\text{r, Earth}}}{A} \Rightarrow F_{\text{r, Earth}} = P_{\text{r, Earth}}A$$

where A is the cross-sectional area of Earth.

Express the radiation pressure in terms of the intensity of the radiation I from the Sun:

$$P_{\text{r, Earth}} = \frac{I}{c}$$

Substituting for $P_{\text{r, Earth}}$ and A yields:

$$F_{\text{r, Earth}} = \frac{I\pi R^2}{c}$$

Substitute numerical values and evaluate F_r:

$$F_{r,\,Earth} = \frac{\pi\left(1.37\,kW/m^2\right)\left(6.37\times10^6\,m\right)^2}{2.998\times10^8\,m/s}$$

$$= 5.825\times10^8\,N$$

$$= \boxed{5.83\times10^8\,N}$$

The gravitational force exerted on Earth by the Sun is given by:

$$F_{g,\,Earth} = \frac{Gm_{sun}\,m_{earth}}{r^2}$$

where r is the radius of Earth's orbit.

Substitute numerical values and evaluate $F_{g,\,Earth}$:

$$F_{g,\,Earth} = \frac{\left(6.673\times10^{-11}\,N\cdot m^2/kg^2\right)\left(1.99\times10^{30}\,kg\right)\left(5.98\times10^{24}\,kg\right)}{\left(1.50\times10^{11}\,m\right)^2} = 3.529\times10^{22}\,N$$

Express the ratio of the force due to radiation pressure $F_{r,\,Earth}$ to the gravitational force $F_{g,\,Earth}$:

$$\frac{F_{r,\,Earth}}{F_{g,\,Earth}} = \frac{5.825\times10^8\,N}{3.529\times10^{22}\,N} = 1.65\times10^{-14}$$

or

$$F_{r,\,Earth} = \boxed{\left(1.65\times10^{-14}\right)F_{g,\,Earth}}$$

(b) The radiation pressure exerted on Mars is given by:

$$P_{r,\,Mars} = \frac{F_{r,\,Mars}}{A} \Rightarrow F_{r,\,Mars} = P_{r,\,Mars}A$$

where A is the cross-sectional area of Mars.

Express the radiation pressure on Mars in terms of the intensity of the radiation I_{Mars} from the sun:

$$P_{r,\,Mars} = \frac{I_{Mars}}{c}$$

Substituting for $P_{r,\,Mars}$ and A yields:

$$F_{r,\,Mars} = \frac{I_{Mars}\,\pi\,R_{Mars}^2}{c}$$

Express the ratio of the solar constant at Earth to the solar constant at Mars:

$$\frac{I_{Mars}}{I_{earth}} = \left(\frac{r_{earth}}{r_{Mars}}\right)^2 \Rightarrow I_{Mars} = I_{earth}\left(\frac{r_{earth}}{r_{Mars}}\right)^2$$

Substitute for I_{Mars} to obtain:

$$F_{r,\,Mars} = \frac{I_{earth}\,\pi\,R_{Mars}^2}{c}\left(\frac{r_{earth}}{r_{Mars}}\right)^2$$

Substitute numerical values and evaluate $F_{\text{r, Mars}}$:

$$F_{\text{r, Mars}} = \frac{\pi\left(1.37\,\text{kW/m}^2\right)\left(3.40\times10^3\ \text{km}\right)^2}{2.998\times10^8\ \text{m/s}}\left(\frac{1.50\times10^{11}\ \text{m}}{2.28\times10^{11}\ \text{m}}\right)^2 = \boxed{7.18\times10^7\ \text{N}}$$

The gravitational force exerted on Mars by the Sun is given by:

$$F_{\text{g, Mars}} = \frac{Gm_{\text{sun}}m_{\text{Mars}}}{r^2} = \frac{Gm_{\text{sun}}\left(0.11m_{\text{Earth}}\right)}{r^2}$$

where r is the radius of Mars' orbit.

Substitute numerical values and evaluate F_{g}

$$F'_{\text{g, Mars}} = \frac{\left(6.673\times10^{-11}\ \text{N}\cdot\text{m}^2/\text{kg}^2\right)\left(1.99\times10^{30}\ \text{kg}\right)\left(0.11\right)\left(5.98\times10^{24}\ \text{kg}\right)}{\left(2.28\times10^{11}\ \text{m}\right)^2}$$

$$= 1.68\times10^{21}\ \text{N}$$

Express the ratio of the force due to radiation pressure $F_{\text{r, Mars}}$ to the gravitational force $F_{\text{g, Mars}}$:

$$\frac{F_{\text{r, Mars}}}{F_{\text{g, Mars}}} = \frac{7.18\times10^7\ \text{N}}{1.68\times10^{21}\ \text{N}} = 4.27\times10^{-14}$$

or

$$F_{\text{r, Mars}} = \boxed{\left(4.27\times10^{-14}\right)F_{\text{g, Mars}}}$$

(c) Because the ratio of the radiation pressure force to the gravitational force is 1.65×10^{-14} for Earth and 4.27×10^{-14} for Mars, Mars has the larger ratio. The reason that the ratio is higher for Mars is that the dependence of the radiation pressure on the distance from the Sun is the same for both forces (r^{-2}), whereas the dependence on the radii of the planets is different. Radiation pressure varies as R^2, whereas the gravitational force varies as R^3 (assuming that the two planets have the same density, an assumption that is nearly true). Consequently, the ratio of the forces goes as $R^2/R^3 = R^{-1}$. Because Mars is smaller than Earth, the ratio is larger.

The Wave Equation for Electromagnetic Waves

45 ••   Show that any function of the form $y(x,\,t) = f(x-vt)$ or $y(x,\,t) = g(x+vt)$ satisfies the wave Equation 30-7

Picture the Problem We can show that these functions satisfy the wave equations by differentiating them twice (using the chain rule) with respect to x and t and equating the expressions for the second partial of f with respect to u.

Let $u = x - vt$. Then:

$$\frac{\partial f}{\partial x} = \frac{\partial u}{\partial x}\frac{\partial f}{\partial u} = \frac{\partial f}{\partial u}$$

and

$$\frac{\partial f}{\partial t} = \frac{\partial u}{\partial t}\frac{\partial f}{\partial u} = -v\frac{\partial f}{\partial u}$$

Express the second derivatives of f with respect to x and t to obtain:

$$\frac{\partial^2 f}{\partial x^2} = \frac{\partial^2 f}{\partial u^2} \text{ and } \frac{\partial^2 f}{\partial t^2} = v^2\frac{\partial^2 f}{\partial u^2}$$

Divide the first of these equations by the second to obtain:

$$\frac{\dfrac{\partial^2 f}{\partial x^2}}{\dfrac{\partial^2 f}{\partial t^2}} = \frac{1}{v^2} \Rightarrow \boxed{\frac{\partial^2 f}{\partial x^2} = \frac{1}{v^2}\frac{\partial^2 f}{\partial t^2}}$$

Let $u = x + vt$. Then:

$$\frac{\partial f}{\partial x} = \frac{\partial u}{\partial x}\frac{\partial f}{\partial u} = \frac{\partial f}{\partial u}$$

and

$$\frac{\partial f}{\partial t} = \frac{\partial u}{\partial t}\frac{\partial f}{\partial u} = v\frac{\partial f}{\partial u}$$

Express the second derivatives of f with respect to x and t to obtain:

$$\frac{\partial^2 f}{\partial x^2} = \frac{\partial^2 f}{\partial u^2} \text{ and } \frac{\partial^2 f}{\partial t^2} = v^2\frac{\partial^2 f}{\partial u^2}$$

Divide the first of these equations by the second to obtain:

$$\frac{\dfrac{\partial^2 f}{\partial x^2}}{\dfrac{\partial^2 f}{\partial t^2}} = \frac{1}{v^2} \Rightarrow \boxed{\frac{\partial^2 f}{\partial x^2} = \frac{1}{v^2}\frac{\partial^2 f}{\partial t^2}}$$

General Problems

47 •• A circular loop of wire can be used to detect electromagnetic waves. Suppose the signal strength from a 100-MHz FM radio station 100 km distant is 4.0 μW/m^2, and suppose the signal is vertically polarized. What is the maximum rms voltage induced in your antenna, assuming your antenna is a 10.0-cm-radius loop?

Picture the Problem We can use Faraday's law to show that the maximum rms voltage induced in the loop is given by $\mathcal{E}_{\text{rms}} = A\omega B_0 /\sqrt{2}$, where A is the area of the loop, B_0 is the amplitude of the magnetic field, and ω is the angular frequency of the wave. Relating the intensity of the radiation to B_0 will allow us to express

$\mathcal{E}_{rms}$ as a function of the intensity.

The emf induced in the antenna is given by Faraday's law:

$$\mathcal{E} = -\frac{d\phi_m}{dt} = -\frac{d}{dt}\left(\vec{B} \cdot A\hat{n}\right) = -\frac{d}{dt}(BA)$$

$$= -A\frac{dB}{dt} = -\pi R^2 \frac{d}{dt}\left(B_0 \sin \omega t\right)$$

$$= -\pi R^2 \omega B_0 \cos \omega t = -\mathcal{E}_{peak} \cos \omega t$$

where $\mathcal{E}_{peak} = \pi R^2 \omega B_0$ and R is the radius of the loop antenna..

$\mathcal{E}_{rms}$ equals $\mathcal{E}_{peak}$ divided by the square root of 2:

$$\mathcal{E}_{rms} - \frac{\mathcal{E}_{peak}}{\sqrt{2}} - \frac{\pi R^2 \omega B_0}{\sqrt{2}} \qquad (1)$$

The intensity of the signal is given by:

$$I = \frac{E_0 B_0}{2\mu_0}$$

or, because $E_0 = cB_0$,

$$I = \frac{cB_0 B_0}{2\mu_0} = \frac{B_0^2 c}{2\mu_0}$$

Solving for B_0 yields:

$$B_0 = \sqrt{\frac{2\mu_0 I}{c}}$$

Substituting for B_0 and ω in equation (1) and simplifying yields:

$$\mathcal{E}_{rms} = \frac{\pi R^2 (2\pi f)\sqrt{\dfrac{2\mu_0 I}{c}}}{\sqrt{2}}$$

$$= 2\pi^2 R^2 f \sqrt{\frac{\mu_0 I}{c}}$$

Substitute numerical values and evaluate $\mathcal{E}_{rms}$:

$$\mathcal{E}_{rms} = 2\pi^2 (0.100\,\text{m})^2 (100\,\text{MHz})\sqrt{\frac{\left(4\pi \times 10^{-7}\,\text{N/A}^2\right)\left(4.0\,\mu\text{W/m}^2\right)}{2.998 \times 10^8\,\text{m/s}}} = \boxed{2.6\,\text{mV}}$$

51 •• The electric fields of two harmonic electromagnetic waves of angular frequency ω_1 and ω_2 are given by $\vec{E}_1 = E_{1,0}\cos\left(k_1 x - \omega_1 t\right)\hat{j}$ and by $\vec{E}_2 = E_{2,0}\cos\left(k_2 x - \omega_2 t + \delta\right)\hat{j}$. For the resultant of these two waves, find (*a*) the instantaneous Poynting vector and (*b*) the time-averaged Poynting vector.

(c) Repeat Parts (a) and (b) if the direction of propagation of the second wave is reversed so that $\vec{E}_2 = E_{2,0}\cos(k_2 x + \omega_2 t + \delta)\hat{j}$

Picture the Problem We can use the definition of the Poynting vector and the relationship between $\vec{B}$ and $\vec{E}$ to find the instantaneous Poynting vectors for each of the resultant wave motions and the fact that the time average of the cross product term is zero for $\omega_1 \neq \omega_2$, and ½ for the square of cosine function to find the time-averaged Poynting vectors.

(a) Because both waves propagate in the x direction:

$$\vec{E} \times \vec{B} = \mu_0 S \hat{i} \Rightarrow \vec{B} = B\hat{k}$$

Express B in terms of E_1 and E_2:

$$B = \frac{1}{c}(E_1 + E_2)$$

Substitute for E_1 and E_2 to obtain:

$$\vec{B}(x,t) = \frac{1}{c}\left[E_{1,0}\cos(k_1 x - \omega_1 t) + E_{2,0}\cos(k_2 x - \omega_2 t + \delta)\right]\hat{k}$$

The instantaneous Poynting vector for the resultant wave motion is given by:

$$\vec{S}(x,t) = \frac{1}{\mu_0}\left(E_{1,0}\cos(k_1 x - \omega_1 t) + E_{2,0}\cos(k_2 x - \omega_2 t + \delta)\right)\hat{j}$$

$$\times \frac{1}{c}\left(E_{1,0}\cos(k_1 x - \omega_1 t) + E_{2,0}\cos(k_2 x - \omega_2 t + \delta)\right)\hat{k}$$

$$= \frac{1}{\mu_0 c}\left(E_{1,0}\cos(k_1 x - \omega_1 t) + E_{2,0}\cos(k_2 x - \omega_2 t + \delta)\right)^2\left(\hat{j} \times \hat{k}\right)$$

$$= \boxed{\frac{1}{\mu_0 c}\left[E_{1,0}^2\cos^2(k_1 x - \omega_1 t) + 2E_{1,0}E_{2,0}\cos(k_1 x - \omega_1 t) \\ \times\cos(k_2 x - \omega_2 t + \delta) + E_{2,0}^2\cos^2(k_2 x - \omega_2 t + \delta)\right]\hat{i}}$$

(b) The time average of the cross product term is zero for $\omega_1 \neq \omega_2$, and the time average of the square of the cosine terms is ½:

$$\vec{S}_{av} = \boxed{\frac{1}{2\mu_0 c}\left[E_{1,0}^2 + E_{2,0}^2\right]\hat{i}}$$

(c) In this case $\vec{B}_2 = -B\hat{k}$ because the wave with $k = k_2$ propagates in the $-\hat{i}$ direction. The magnetic field is then:

$$\vec{B}(x,t)=\frac{1}{c}\Big[E_{1,0}\cos(k_1 x-\omega_1 t)-E_{2,0}\cos(k_2 x+\omega_2 t+\delta)\Big]\hat{k}$$

The instantaneous Poynting vector for the resultant wave motion is given by:

$$\vec{S}(x,t)=\frac{1}{\mu_0}\big(E_{1,0}\cos(k_1 x-\omega_1 t)+E_{2,0}\cos(k_2 x-\omega_2 t+\delta)\big)\hat{j}$$

$$\times\frac{1}{c}\big(E_{1,0}\cos(k_1 x-\omega_1 t)-E_{2,0}\cos(k_2 x+\omega_2 t+\delta)\big)\hat{k}$$

$$=\boxed{\frac{1}{\mu_0 c}\Big[E_{1,0}^2\cos^2(k_1 x-\omega_1 t)-E_{2,0}^2\cos^2(k_2 x+\omega_2 t+\delta)\Big]\hat{i}}$$

The time average of the square of the cosine terms is ½:

$$\vec{S}_{av}=\boxed{\frac{1}{2\mu_0 c}\Big[E_{1,0}^2-E_{2,0}^2\Big]\hat{i}}$$

55 ••• A conductor in the shape of a long solid cylinder that has a length L, a radius a, and a resistivity ρ carries a steady current I that is uniformly distributed over its cross-section. (*a*) Use Ohm's law to relate the electric field $\vec{E}$ in the conductor to I, ρ, and a. (*b*) Find the magnetic field $\vec{B}$ just outside the conductor. (*c*) Use the results from Part (*a*) and Part (*b*) to compute the Poynting vector $\vec{S}=(\vec{E}\times\vec{B})/\mu_0$ at $r=a$ (the edge of the conductor). In what direction is $\vec{S}$? (*d*) Find the flux $\oint S_n dA$ through the surface of the cylinder, and use this flux to show that the rate of energy flow into the conductor equals $I^2 R$, where R is the resistance of the cylinder.

Picture the Problem A side view of the cylindrical conductor is shown in the diagram. Let the current be to the right (in the +x direction) and choose a coordinate system in which the +y direction is radially outward from the axis of the conductor. Then the +z direction is tangent to cylindrical surfaces that are concentric with the axis of the conductor (out of the plane of the diagram at the location indicated in the diagram). We can use Ohm's law to relate the electric field strength E in the conductor to I, ρ, and a and Ampere's law to find the magnetic field strength B just outside the conductor. Knowing $\vec{E}$ and $\vec{B}$ we can find $\vec{S}$ and, using its normal component, show that the rate of energy flow into the conductor equals $I^2 R$, where R is the resistance.

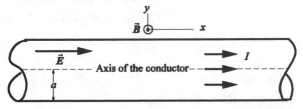

(a) Apply Ohm's law to the cylindrical conductor to obtain:

$$V = IR = \frac{I\rho L}{A} = \frac{I\rho}{\pi a^2} L = EL$$

where $E = \frac{I\rho}{\pi a^2}$.

Because $\vec{E}$ is in the same direction as I:

$$\vec{E} = \boxed{\frac{I\rho}{\pi a^2} \hat{i}} \text{ where } \hat{i} \text{ is a unit vector}$$

in the direction of the current.

(b) Applying Ampere's law to a circular path of radius a at the surface of the cylindrical conductor yields:

$$\oint_C \vec{B} \cdot d\vec{\ell} = B(2\pi a) = \mu_0 I_{enclosed} = \mu_0 I$$

Solve for the magnetic field strength B to obtain:

$$B = \frac{\mu_0 I}{2\pi a}$$

Apply a right-hand rule to determine the direction of $\vec{B}$ at the point of interest shown in the diagram:

$$\vec{B} = \boxed{\frac{\mu_0 I}{2\pi a} \hat{\theta}} \text{ where } \hat{\theta} \text{ is a unit vector}$$

perpendicular to $\hat{i}$ and tangent to the surface of the conducting cylinder.

(c) The Poynting vector is given by:

$$\vec{S} = \frac{1}{\mu_0} \vec{E} \times \vec{B}$$

Substitute for $\vec{E}$ and $\vec{B}$ and simplify to obtain:

$$\vec{S} = \frac{1}{\mu_0} \left(\frac{I\rho}{\pi a^2} \right) \hat{i} \times \left(\frac{\mu_0 I}{2\pi a} \right) \hat{k}$$

$$= -\frac{I^2 \rho}{2\pi^2 a^3} \hat{j}$$

Letting $\hat{r}$ be a unit vector directed radially outward from the axis of the cylindrical conductor yields.

$$\vec{S} = \boxed{-\frac{I^2 \rho}{2\pi^2 a^3} \hat{r}} \text{ where } \hat{r} \text{ is a unit}$$

vector directed radially outward away from the axis of the conducting cylinder.

(d) The flux through the surface of the conductor into the conductor is:

$$\oint S_n dA = S(2\pi a L)$$

Substitute for S_n, the *inward* component of $\vec{S}$, and simplify to obtain:

$$\oint S_n \, dA = \frac{I^2 \rho}{2\pi^2 a^3}(2\pi a L) = \frac{I^2 \rho L}{\pi a^2}$$

Because $R = \dfrac{\rho L}{A} = \dfrac{\rho L}{\pi a^2}$:

$$\oint S_n \, dA = \boxed{I^2 R}$$

Remarks: The equality of the two flow rates is a statement of the conservation of energy.

59 ••• An intense point source of light radiates 1.00 MW isotropically (uniformly in all directions). The source is located 1.00 m above an infinite, perfectly reflecting plane. Determine the force that the radiation pressure exerts on the plane.

Picture the Problem Let the point source be a distance a above the plane. Consider a ring of radius r and thickness dr in the plane and centered at the point directly below the light source. Express the force on this elemental ring and integrate the resulting expression to obtain F.

The intensity anywhere along this infinitesimal ring is given by:

$$\frac{P}{4\pi\left(r^2 + a^2\right)}$$

The elemental force dF on the elemental ring of area $2\pi r \, dr$ is given by:

$$dF = \frac{P\,r\,dr}{c\left(r^2 + a^2\right)}\frac{a}{\sqrt{r^2 + a^2}}$$

$$= \frac{P a r \, dr}{c\left(r^2 + a^2\right)^{3/2}}$$

where we have taken into account that only the normal component of the incident radiation contributes to the force on the plane, and that the plane is a perfectly reflecting plane.

Integrate dF from $r = 0$ to $r = \infty$:

$$F = \frac{P a}{c}\int_0^\infty \frac{r\,dr}{\left(r^2 + a^2\right)^{3/2}}$$

From integral tables:

$$\int_0^\infty \frac{r\,dr}{\left(r^2 + a^2\right)^{3/2}} = \frac{-1}{\sqrt{r^2 + a^2}}\Bigg]_0^\infty = \frac{1}{a}$$

Substitute to obtain:

$$F = \frac{Pa}{c}\left(\frac{1}{a}\right) = \frac{P}{c}$$

Substitute numerical values and evaluate F:

$$F = \frac{1.00\,\text{MW}}{2.998 \times 10^8\,\text{m/s}} = \boxed{3.34\,\text{mN}}$$

Chapter 31
Properties of Light

Conceptual Problems

1 • A ray of light reflects from a plane mirror. The angle between the incoming ray and the reflected ray is 70°. What is the angle of reflection? (*a*) 70°, (*b*) 140°, (*c*) 35°, (*d*) Not enough information is given to determine the reflection angle.

Determine the Concept Because the angles of incidence and reflection are equal, their sum is 70° and the angle of reflection is 35°. $\boxed{(c)}$ is correct.

7 • A swimmer at point *S* in Figure 31-53 develops a leg cramp while swimming near the shore of a calm lake and calls for help. A lifeguard at point *L* hears the call. The lifeguard can run 9.0 m/s and swim 3.0 m/s. She knows physics and chooses a path that will take the least time to reach the swimmer. Which of the paths shown in the figure does the lifeguard take?

Determine the Concept The path through point *D* is the path of least time. In analogy to the refraction of light, the ratio of the sine of the angle of incidence to the sine of the angle of refraction equals the ratio of the speeds of the lifeguard in each medium. Careful measurements from the figure show that path *LDS* is the path that best satisfies this criterion.

9 • A human eye perceives color using a structure which is called a *cone* that is is located on the retina. Three types of molecules compose these cones and each type of molecule absorbs either red, green, or blue light by resonance absorption. Use this fact to explain why the color of an object that appears blue in air appears blue underwater, in spite of the fact that the wavelength of the light is shortened in accordance with Equation 31-6.

Determine the Concept In resonance absorption, the molecules respond to the frequency of the light through the Einstein photon relation $E = hf$. Neither the wavelength nor the frequency of the light within the eyeball depend on the index of refraction of the medium outside the eyeball. Thus, the color appears to be the same in spite of the fact that the wavelength has changed.

11 •• Draw a diagram to explain how Polaroid sunglasses reduce glare from sunlight reflected from a smooth horizontal surface, such as the surface found on a pool of water. Your diagram should clearly indicate the direction of polarization of the light as it propagates from the Sun to the reflecting surface and then through the sunglasses into the eye.

Determine the Concept The following diagram shows unpolarized light from the sun incident on the smooth surface at the polarizing angle for that particular surface. The reflected light is polarized perpendicular to the plane of incidence, i.e., in the horizontal direction. The sunglasses are shown in the correct orientation to pass vertically polarized light and block the reflected sunlight.

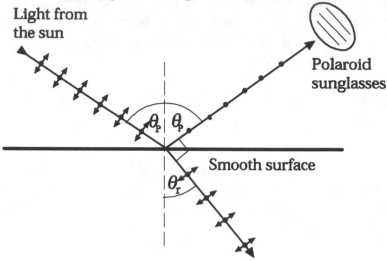

15 •   What role does the helium play in a helium–neon laser?

Determine the Concept The population inversion between the state $E_{2,Ne}$ and the state 1.96 eV below it (see Figure 31-51) is achieved by inelastic collisions between neon atoms and helium atoms excited to the state $E_{2,He}$.

17 •   Which of the following types of light would have the highest energy photons? (*a*) red (*b*) infrared (*c*) blue (*d*) ultraviolet

Determine the Concept The energy of a photon is directly proportional to the frequency of the light and inversely proportional to its wavelength. Of the portions of the electromagnetic spectrum include in the list of answers, ultraviolet light has the highest frequency. $\boxed{(d)}$ is correct.

The Speed of Light

25 ••   Ole Römer discovered the finiteness of the speed of light by observing Jupiter's moons. Approximately how sensitive would the timing apparatus need to be in order to detect a shift in the predicted time of the moon's eclipses that occur when the moon happens to be at perigee (3.63×10^5 km) and those that occur when the moon is at apogee (4.06×10^5 km)? Assume that an instrument should be able to measure to at least one-tenth the magnitude of the effect it is to measure.

Picture the Problem His timing apparatus would need to be sensitive enough to measure the difference in times for light to travel to Earth when the moon is at perigee and at apogee.

The sensitivity of the timing apparatus would need to be one-tenth of the difference in time for light to reach Earth from the two positions of the moon of Jupiter:

$$\text{Sensitivity} = \tfrac{1}{10}\Delta t$$

where Δt is the time required for light to travel between the two positions of the moon.

The time required for light to travel between the two positions of the moon is given by:

$$\Delta t = \frac{d_{\text{moon at apogee}} - d_{\text{moon at perigee}}}{c}$$

Substituting for Δt yields:

$$\text{Sensitivity} = \frac{d_{\text{moon at apogee}} - d_{\text{moon at perigee}}}{10c}$$

Substitute numerical values and evaluate the required sensitivity:

$$\Delta t = \frac{4.06\times10^5 \text{ km} - 3.63\times10^5 \text{ km}}{(10)(2.998\times10^8 \text{ m/s})}$$

$$= \boxed{14 \text{ ms}}$$

Remarks: Instruments with this sensitivity did not exist in the 17ᵗʰ century.

Reflection and Refraction

31 •• A slab of glass that has an index of refraction of 1.50 is submerged in water that has an index of refraction of 1.33. Light in the water is incident on the glass. Find the angle of refraction if the angle of incidence is (a) 60°, (b) 45°, and (c) 30°.

Picture the Problem Let the subscript 1 refer to the water and the subscript 2 to the glass and apply Snell's law to the water-glass interface.

Apply Snell's law to the water-glass interface to obtain:

$$n_1 \sin\theta_1 = n_2 \sin\theta_2$$

Solving for θ_2 yields:

$$\theta_2 = \sin^{-1}\!\left(\frac{n_1}{n_2}\sin\theta_1\right)$$

(a) Evaluate θ_2 for $\theta_1 = 60°$:

$$\theta_2 = \sin^{-1}\!\left(\frac{1.33}{1.50}\sin 60°\right) = \boxed{50°}$$

(b) Evaluate θ_2 for $\theta_1 = 45°$:

$$\theta_2 = \sin^{-1}\left(\frac{1.33}{1.50}\sin 45°\right) = \boxed{39°}$$

(c) Evaluate θ_2 for $\theta_1 = 30°$:

$$\theta_2 = \sin^{-1}\left(\frac{1.33}{1.50}\sin 30°\right) = \boxed{26°}$$

35 •• In Figure 31-56, light is initially in a medium that has an index of refraction n_1. It is incident at angle θ_1 on the surface of a liquid that has an index of refraction n_2. The light passes through the layer of liquid and enters glass that has an index of refraction n_3. If θ_3 is the angle of refraction in the glass, show that $n_1 \sin \theta_1 = n_3 \sin \theta_3$. That is, show that the second medium can be neglected when finding the angle of refraction in the third medium.

Picture the Problem We can apply Snell's law consecutively, first to the n_1-n_2 interface and then to the n_2-n_3 interface.

Apply Snell's law to the n_1-n_2 interface:

$$n_1 \sin\theta_1 = n_2 \sin\theta_2$$

Apply Snell's law to the n_2-n_3 interface:

$$n_2 \sin\theta_2 = n_3 \sin\theta_3$$

Equate the two expressions for $n_2 \sin\theta_2$ to obtain:

$$\boxed{n_1 \sin\theta_1 = n_3 \sin\theta_3}$$

Total Internal Reflection

39 • What is the critical angle for light traveling in water that is incident on a water–air interface?

Picture the Problem Let the subscript 1 refer to the water and the subscript 2 to the air and use Snell's law under total internal reflection conditions.

Use Snell's law to obtain:

$$n_1 \sin\theta_1 = n_2 \sin\theta_2$$

When there is total internal reflection:

$$\theta_1 = \theta_c \text{ and } \theta_2 = 90°$$

Substitute to obtain:

$$n_1 \sin\theta_c = n_2 \sin 90° = n_2$$

Solving for θ_c yields:

$$\theta_c = \sin^{-1}\left(\frac{n_2}{n_1}\right)$$

| Substitute numerical values and evaluate θ_c: | $\theta_c = \sin^{-1}\left(\dfrac{1.00}{1.33}\right) = \boxed{48.8°}$ |

45 ••   Find the maximum angle of incidence θ_1 of a ray that would propagate through an optical fiber that has a core index of refraction of 1.492, a core radius of 50.00 μm, and a cladding index of 1.489. See Problem 44.

Picture the Problem We can use the result of Problem 44 to find the maximum angle of incidence under the given conditions.

| From Problem 44: | $\sin\theta_1 = \sqrt{n_2^2 - n_3^2}$ |

| Solve for θ_1 to obtain: | $\theta_1 = \sin^{-1}\left(\sqrt{n_2^2 - n_3^2}\right)$ |

| Substitute numerical values and evaluate θ_1: | $\theta_1 = \sin^{-1}\left(\sqrt{(1.492)^2 - (1.489)^2}\right)$ |
| | $= \boxed{5°}$ |

Polarization

51 •   What is the polarizing angle for light in air that is incident on (a) water ($n = 1.33$), and (b) glass ($n = 1.50$)?

Picture the Problem The polarizing angle is given by Brewster's law: $\tan\theta_p = n_2/n_1$ where n_1 and n_2 are the indices of refraction on the near and far sides of the interface, respectively.

| Use Brewster's law to obtain: | $\theta_p = \tan^{-1}\left(\dfrac{n_2}{n_1}\right)$ |

| (a) For $n_1 = 1$ and $n_2 = 1.33$: | $\theta_p = \tan^{-1}\left(\dfrac{1.33}{1.00}\right) = \boxed{53.1°}$ |

| (b) For $n_1 = 1$ and $n_2 = 1.50$: | $\theta_p = \tan^{-1}\left(\dfrac{1.50}{1.00}\right) = \boxed{56.3°}$ |

55 ••   The polarizing angle for light in air that is incident on a certain substance is 60°. (a) What is the angle of refraction of light incident at this angle? (b) What is the index of refraction of this substance?

Picture the Problem Assume that light is incident in air ($n_1 = 1.00$). We can use the relationship between the polarizing angle and the angle of refraction to determine the latter and Brewster's law to find the index of refraction of the substance.

(a) At the polarizing angle, the sum of the angles of polarization and refraction is 90°:

$$\theta_p + \theta_r = 90° \Rightarrow \theta_r = 90° - \theta_p$$

Substitute for θ_p to obtain:

$$\theta_r = 90° - 60° = \boxed{30°}$$

(b) From Brewster's law we have:

$$\tan\theta_p = \frac{n_2}{n_1}$$

or, because $n_1 = 1.00$,
$$n_2 = \tan\theta_p$$

Substitute for θ_p and evaluate n_2:

$$n_2 = \tan 60° = \boxed{1.7}$$

59 ••   The device described in Problem 58 could serve as a *polarization rotator*, which changes the linear plane of polarization from one direction to another. The efficiency of such a device is measured by taking the ratio of the output intensity at the desired polarization to the input intensity. The result of Problem 58 suggests that the highest efficiency is achieved by using a large value for the number N. A small amount of intensity is lost regardless of the input polarization when using a real polarizer. For each polarizer, assume the transmitted intensity is 98 percent of the amount predicted by the law of Malus and use a **spreadsheet** or graphing program to determine the optimum number of sheets you should use to rotate the polarization 90°.

Picture the Problem Let I_n be the intensity after the nth polarizing sheet and use $I = I_0 \cos^2\theta$ to find the ratio of I_{n+1} to I_n. Because each sheet introduces a 2% loss of intensity, the net transmission after N sheets $(0.98)^N$.

Find the ratio of I_{n+1} to I_n:

$$\frac{I_{n+1}}{I_n} = (0.98)\cos^2\frac{\pi}{2N}$$

Because there are N such reductions of intensity:

$$\frac{I_{N+1}}{I_0} = (0.98)^N \cos^{2N}\left(\frac{\pi}{2N}\right)$$

A spreadsheet program to graph I_{N+1}/I_0 for an ideal polarizer as a function of N, the percent transmission, and I_{N+1}/I_0 for a real polarizer as a function of N is shown below. The formulas used to calculate the quantities in the columns are as follows:

Cell	Content/Formula	Algebraic Form
A3	1	N
B2	(cos(PI()/(2*A2))^(2*A2)	$\cos^{2N}\left(\dfrac{\pi}{2N}\right)$
C3	(0.98)^A3	$(0.98)^N$
D4	B3*C3	$(0.98)^N \cos^{2N}\left(\dfrac{\pi}{2N}\right)$

	A	B	C	D
		Ideal	Percent	Real
2	N	Polarizer	Transmission	Polarizer
3	1	0.000	0.980	0.000
4	2	0.250	0.960	0.240
5	3	0.422	0.941	0.397
6	4	0.531	0.922	0.490
7	5	0.605	0.904	0.547
8	6	0.660	0.886	0.584
9	7	0.701	0.868	0.608
10	8	0.733	0.851	0.624
11	9	0.759	0.834	0.633
12	10	0.781	0.817	0.638
13	11	0.798	0.801	0.639
14	12	0.814	0.785	0.638
15	13	0.827	0.769	0.636
16	14	0.838	0.754	0.632
17	15	0.848	0.739	0.626
18	16	0.857	0.724	0.620
19	17	0.865	0.709	0.613
20	18	0.872	0.695	0.606
21	19	0.878	0.681	0.598
22	20	0.884	0.668	0.590

A graph of I/I_0 as a function of N for the quantities described above follows:

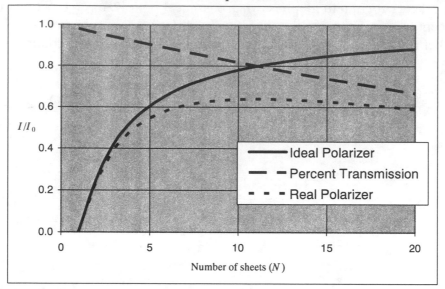

Inspection of the table, as well as of the graph, tells us that the optimum number of sheets is   11.

63 ••   A circularly polarized wave is said to be *right circularly polarized* if the electric and magnetic fields rotate clockwise when viewed along the direction of propagation and *left circularly polarized* if the fields rotate counterclockwise. (*a*) What is the sense of the circular polarization for the wave described by the expression in Problem 62? (*b*) What would be the expression for the electric field of a circularly polarized wave traveling in the same direction as the wave in Problem 60, but with the fields rotating in the opposite direction?

Picture the Problem We can apply the given definitions of right and left circular polarization to the electric field and magnetic fields of the wave.

(*a*) The electric field of the wave in Problem 62 is:

$$\vec{E} = E_0 \sin(kx - \omega t)\hat{j} + E_0 \cos(kx - \omega t)\hat{k}$$

The corresponding magnetic field is:

$$\vec{B} = B_0 \sin(kx - \omega t)\hat{k} - B_0 \cos(kx - \omega t)\hat{j}$$

Because these fields rotate clockwise when viewed along the direction of propagation, the wave is right circularly polarized.

(b) For a left circularly polarized wave traveling in the opposite direction:

$$\vec{E} = \boxed{E_0 \sin(kx + \omega t)\hat{j} - E_0 \cos(kx + \omega t)\hat{k}}$$

Sources of Light

67 ••   Sodium has excited states 2.11 eV, 3.20 eV, and 4.35 eV above the ground state. Assume that the atoms of the gas are all in the ground state prior to irradiation. (a) What is the maximum wavelength of radiation that will result in resonance fluorescence? What is the wavelength of the fluorescent radiation? (b) What wavelength will result in excitation of the state 4.35 eV above the ground state? If that state is excited, what are the possible wavelengths of resonance fluorescence that might be observed?

Picture the Problem The ground state and the three excited energy levels are shown in the diagram to the right. Because the wavelength is related to the energy of a photon by $\lambda = hc/\Delta E$, longer wavelengths correspond to smaller energy differences.

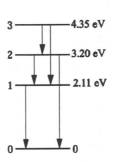

(a) The maximum wavelength of radiation that will result in resonance fluorescence corresponds to an excitation to the 3.20 eV level followed by decays to the 2.11 eV level and the ground state:

$$\lambda_{max} = \frac{1240\,\text{eV}\cdot\text{nm}}{3.20\,\text{eV}} = \boxed{388\,\text{nm}}$$

The fluorescence wavelengths are:

$$\lambda_{2\to1} = \frac{1240\,\text{eV}\cdot\text{nm}}{3.20\,\text{eV} - 2.11\,\text{eV}} = \boxed{1140\,\text{nm}}$$

and

$$\lambda_{1\to0} = \frac{1240\,\text{eV}\cdot\text{nm}}{2.11\,\text{eV} - 0} = \boxed{588\,\text{nm}}$$

(b) For excitation:

$$\lambda_{0\to3} = \frac{1240\,\text{eV}\cdot\text{nm}}{4.35\,\text{eV}} = 285\,\text{nm}$$

(not in visible the spectrum)

The fluorescence wavelengths corresponding to the possible transitions are:

$$\lambda_{3\to2} = \frac{1240\,\text{eV}\cdot\text{nm}}{4.35\,\text{eV}-3.20\,\text{eV}} = 1080\,\text{nm}$$

(not in visible spectrum)

$$\lambda_{2\to1} = \frac{1240\,\text{eV}\cdot\text{nm}}{3.20\,\text{eV}-2.11\,\text{eV}} = 1140\,\text{nm}$$

(not in visible spectrum)

$$\lambda_{1\to0} = \frac{1240\,\text{eV}\cdot\text{nm}}{2.11\,\text{eV}-0} = \boxed{588\,\text{nm}}$$

$$\lambda_{3\to1} = \frac{1240\,\text{eV}\cdot\text{nm}}{4.35\,\text{eV}-2.11\,\text{eV}} = \boxed{554\,\text{nm}}$$

and

$$\lambda_{2\to0} = \frac{1240\,\text{eV}\cdot\text{nm}}{3.20\,\text{eV}-0} = 388\,\text{nm}$$

(not in visible spectrum)

69 •   A pulse from a ruby laser has an average power of 10 MW and lasts 1.5 ns. (*a*) What is the total energy of the pulse? (*b*) How many photons are emitted in this pulse?

Picture the Problem We can use the definition of power to find the total energy of the pulse. The ratio of the total energy to the energy per photon will yield the number of photons emitted in the pulse.

(*a*) Use the definition of power to obtain:

$$E = P\Delta t$$

Substitute numerical values and evaluate E:

$$E = (10\,\text{MW})(1.5\,\text{ns}) = \boxed{15\,\text{mJ}}$$

(*b*) Relate the number of photons N to the total energy in the pulse and the energy of a single photon E_{photon} :

$$N = \frac{E}{E_{\text{photon}}}$$

The energy of a photon is given by:

$$E_{\text{photon}} = \frac{hc}{\lambda}$$

Substitute for E_{photon} to obtain:

$$N = \frac{\lambda E}{hc}$$

Substitute numerical values (the wavelength of light emitted by a ruby laser is 694.3 nm) and evaluate N:

$$N = \frac{(694.3\,\text{nm})(15\,\text{mJ})}{1240\,\text{eV}\cdot\text{nm}} \frac{1\,\text{eV}}{1.602\times10^{-19}\,\text{J}} = \boxed{5.2\times10^{16}}$$

General Problems

71 ••   The critical angle for total internal reflection for a substance is 48°. What is the polarizing angle for this substance?

Picture the Problem We can use Snell's law, under critical angle and polarization conditions, to relate the polarizing angle of the substance to the critical angle for internal reflection.

Apply Snell's law, under critical angle conditions, to the interface:

$$n_1 \sin\theta_c = n_2 \qquad (1)$$

Apply Snell's law, under polarization conditions, to the interface:

$$n_1 \sin\theta_p = n_2 \sin(90° - \theta_p) = n_2 \cos\theta_p$$

or

$$\tan\theta_p = \frac{n_2}{n_1} \Rightarrow \theta_p = \tan^{-1}\left(\frac{n_2}{n_1}\right) \quad (2)$$

Solve equation (1) for the ratio of n_2 to n_1:

$$\frac{n_2}{n_1} = \sin\theta_c$$

Substitute for n_2/n_1 in equation (2) to obtain:

$$\theta_p = \tan^{-1}(\sin\theta_c)$$

Substitute numerical values and evaluate θ_p:

$$\theta_p = \tan^{-1}(\sin 48°) = \boxed{37°}$$

73 ••   Use Figure 31-59 to calculate the critical angles for light initially in silicate flint glass that is incident on a glass–air interface if the light is (*a*) violet light of wavelength 400 nm and (*b*) red light of wavelength 700 nm.

Picture the Problem We can apply Snell's law at the glass-air interface to express θ_c in terms of the index of refraction of the glass and use Figure 31-59 to find the index of refraction of the glass for the given wavelengths of light.

Apply Snell's law at the glass-air interface:

$$n_1 \sin\theta_1 = n_2 \sin\theta_2$$

If $\theta_1 = \theta_c$ and $n_2 = 1$:

$$n_1 \sin \theta_c = \sin 90° = 1$$

and

$$\theta_c = \sin^{-1}\left(\frac{1}{n_1}\right)$$

(a) For violet light of wavelength 400 nm, $n_1 = 1.67$:

$$\theta_c = \sin^{-1}\left(\frac{1}{1.67}\right) = \boxed{36.8°}$$

(b) For red light of wavelength 700 nm, $n_1 = 1.60$:

$$\theta_c = \sin^{-1}\left(\frac{1}{1.60}\right) = \boxed{38.7°}$$

77 ••   From the data provided in Figure 31-59, calculate the polarization angle for an air–glass interface, using light of wavelength 550 nm in each of the four types of glass shown.

Picture the Problem We can use Brewster's law in conjunction with index of refraction data from Figure 31-59 to calculate the polarization angles for the air-glass interface.

From Brewster's law we have:

$$\theta_p = \tan^{-1}\left(\frac{n_2}{n_1}\right)$$

or, for $n_1 = 1$,

$$\theta_p = \tan^{-1}(n_2)$$

For silicate flint glass, $n_2 \approx 1.62$ and:

$$\theta_{p,\text{silicate flint}} = \tan^{-1}(1.62) = \boxed{58.3°}$$

For borate flint glass, $n_2 \approx 1.57$ and:

$$\theta_{p,\text{borate flint}} = \tan^{-1}(1.57) = \boxed{57.5°}$$

For quartz glass, $n_2 \approx 1.54$ and:

$$\theta_{p,\text{quartz}} = \tan^{-1}(1.54) = \boxed{57.0°}$$

For silicate crown glass, $n_2 \approx 1.51$ and:

$$\theta_{p,\text{silicate crown}} = \tan^{-1}(1.51) = \boxed{56.5°}$$

79 ••   (a) For light rays inside a transparent medium that is surrounded by a vacuum, show that the polarizing angle and the critical angle for total internal reflection satisfy $\tan \theta_p = \sin \theta_c$. (b) Which angle is larger, the polarizing angle or the critical angle for total internal reflection?

Picture the Problem We can apply Snell's law at the critical angle and the polarizing angle to show that $\tan \theta_p = \sin \theta_c$.

(*a*) Apply Snell's law at the medium-vacuum interface:

$$n_1 \sin \theta_1 = n_2 \sin \theta_r$$

For $\theta_1 = \theta_c$, $n_1 = n$, and $n_2 = 1$:

$$n \sin \theta_c = \sin 90° = 1$$

For $\theta_1 = \theta_p$, $n_1 = n$, and $n_2 = 1$:

$$\tan \theta_p = \frac{n_2}{n_1} = \frac{1}{n} \Rightarrow n \tan \theta_p = 1$$

Because both expressions equal one:

$$\boxed{\tan \theta_p = \sin \theta_c}$$

(*b*) For any value of θ:

$$\tan \theta > \sin \theta \Rightarrow \boxed{\theta_p > \theta_c}$$

Chapter 32
Optical Images

Conceptual Problems

3 •   True or False

(*a*) The virtual image formed by a concave mirror is always smaller than the object.
(*b*) A concave mirror always forms a virtual image.
(*c*) A convex mirror never forms a real image of a real object.
(*d*) A concave mirror never forms an enlarged real image of an object.

(*a*) False. The size of the virtual image formed by a concave mirror when the object is between the focal point and the vertex of the mirror depends on the distance of the object from the vertex and is always larger than the object.

(*b*) False. When the object is outside the focal point, the image is real.

(*c*) True.

(*d*) False. When the object is between the center of curvature and the focal point, the image is enlarged and real.

5 •   An ant is crawling along the axis of a concave mirror that has a radius of curvature R. At what object distances, if any, will the mirror produce (*a*) an upright image, (*b*) a virtual image, (*c*) an image smaller than the object, and (*d*) an image larger than the object?

Determine the Concept
(*a*) The mirror will produce an upright image for all object distances.

(*b*) The mirror will produce a virtual image for all object distances.

(*c*) The mirror will produce an image that is that is smaller than the object for all object distances.

(*d*) The mirror will never produce an enlarged image.

Plane Mirrors

19 • The image of the object point P in Figure 32-57 is viewed by an eye, as shown. Draw rays from the object point that reflect from the mirror and enter the eye. If the object point and the mirror are fixed in their locations, indicate the range of locations where the eye can be positioned and still see the image of the object point.

Determine the Concept Rays from the source that are reflected by the mirror are shown in the following diagram. The reflected rays appear to diverge from the image. The eye can see the image if it is in the region between rays 1 and 2.

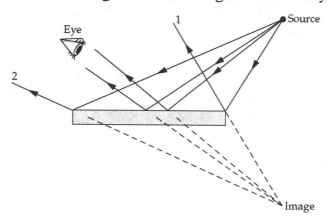

21 •• (a) Two plane mirrors make an angle of 90°. The light from a point object that is arbitrarily positioned in front of the mirrors produces images at three locations. For each image location, draw two rays from the object that, after one or two reflections, appear to come from the image location. (b) Two plane mirrors make an angle of 60° with each other. Draw a sketch to show the location of all the images formed of an object on the bisector of the angle between the mirrors. (c) Repeat Part (b) for an angle of 120°.

Determine the Concept

(a) Draw rays of light from the object (P) that satisfy the law of reflection at the two mirror surfaces. Three virtual images are formed, as shown in the following figure. The eye should be to the right and above the mirrors in order to see these images.

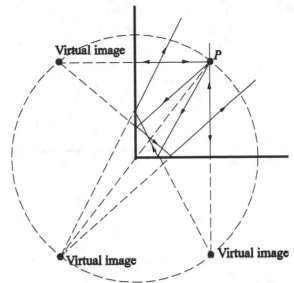

(b) The following diagram shows selected rays emanating from a point object (P) located on the bisector of the 60° angle between the mirrors. These rays form the two virtual images below the horizontal mirror. The construction details for the two virtual images behind the mirror that is at an angle of 60° with the horizontal mirror have been omitted due to the confusing detail their inclusion would add to the diagram.

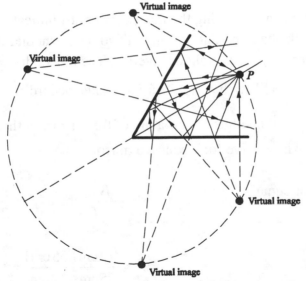

(c) The following diagram shows selected rays emanating from a point object (P) on the bisector of the 120° angle between the mirrors. These rays form the two virtual images at the intersection of the dashed lines (extensions of the reflected rays):

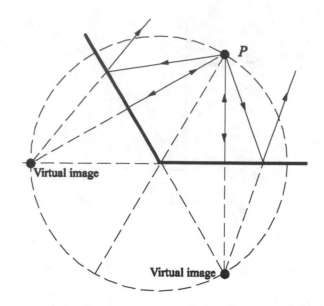

Virtual image

P

Virtual image

Spherical Mirrors

25 •   (*a*) Use the mirror equation (Equation 32-4 where $f = r/2$) to calculate the image distances for the object distances and mirror of Problem 24. (*b*) Calculate the magnification for each given object distance.

Picture the Problem In describing the images, we must indicate where they are located, how large they are in relationship to the object, whether they are real or virtual, and whether they are upright or inverted. The object distance s, the image distance s', and the focal length of a mirror are related according to $\dfrac{1}{s} + \dfrac{1}{s'} = \dfrac{1}{f}$, where $f = \frac{1}{2}r$ and r is the radius of curvature of the mirror. In this problem, $f = 12$ cm because r is positive for a concave mirror.

(*a*) Solve the mirror equation for s':

$$s' = \frac{fs}{s - f} \text{ where } f = \frac{r}{2}$$

When $s = 55$ cm:

$$s' = \frac{(12\,\text{cm})(55\,\text{cm})}{55\,\text{cm} - 12\,\text{cm}} = 15.35\,\text{cm}$$

$$= \boxed{15\,\text{cm}}$$

When $s = 24$ cm:

$$s' = \frac{(12\,\text{cm})(24\,\text{cm})}{24\,\text{cm} - 12\,\text{cm}} = \boxed{24\,\text{cm}}$$

When $s = 12$ cm:

$$s' = \frac{(12\,\text{cm})(12\,\text{cm})}{12\,\text{cm} - 12\,\text{cm}} \boxed{\text{is undefined}}$$

When $s = 8.0$ cm:

$$s' = \frac{(12\,\text{cm})(8.0\,\text{cm})}{8.0\,\text{cm} - 12\,\text{cm}} = -24.0\,\text{cm}$$

$$= \boxed{-0.2\,\text{m}}$$

(b) The lateral magnification of the image is:

$$m = -\frac{s'}{s}$$

When $s = 55$ cm, the lateral magnification of the image is:

$$m = -\frac{s'}{s} = -\frac{15.35\,\text{cm}}{55\,\text{cm}} = \boxed{-0.28}$$

When $s = 24$ cm, the lateral magnification of the image is:

$$m = -\frac{s'}{s} = -\frac{24\,\text{cm}}{24\,\text{cm}} = \boxed{-1.0}$$

When $s = 12$ cm:

$$\boxed{m \text{ is undefined.}}$$

When $s = 8.0$ cm, the lateral magnification of the image is:

$$m = -\frac{s'}{s} = -\frac{-24\,\text{cm}}{8.0\,\text{cm}} = \boxed{3.0}$$

Remarks: These results are in excellent agreement with those obtained graphically in Problem 24.

29 • A dentist wants a small mirror that will produce an upright image that has a magnification of 5.5 when the mirror is located 2.1 cm from a tooth. (a) Should the mirror be concave or convex? (b) What should the radius of curvature of the mirror be?

Picture the Problem We can use the mirror equation and the definition of the lateral magnification to find the radius of curvature of the dentist's mirror.

(a) The mirror must be concave. A convex mirror always produces a diminished virtual image.

(b) Express the mirror equation:

$$\frac{1}{s} + \frac{1}{s'} = \frac{1}{f} = \frac{2}{r} \Rightarrow r = \frac{2ss'}{s'+s} \quad (1)$$

The lateral magnification of the mirror is given by:

$$m = -\frac{s'}{s} \Rightarrow s' = -ms$$

Substitute for s' in equation (1) to obtain:

$$r = \frac{-2ms}{1-m}$$

Substitute numerical values and evaluate r:

$$r = \frac{-2(5.5)(2.1\text{cm})}{1-5.5} = \boxed{5.1\text{cm}}$$

Images Formed by Refraction

35 •   A fish is 10 cm from the front surface of a spherical fish bowl of radius 20 cm. (*a*) How far behind the surface of the bowl does the fish appear to someone viewing the fish from in front of the bowl? (*b*) By what distance does the fish's apparent location change (relative to the front surface of the bowl) when it swims away to 30 cm from the front surface?

Picture the Problem The diagram shows two rays (from the bundle of rays) of light refracted at the water-air interface. Because the index of refraction of air is less than that of water, the rays are bent away from the normal. The fish will, therefore, appear to be closer than it actually is. We can use the equation for refraction at a single surface to find the distance s'. We'll assume that the glass bowl is thin enough that we can ignore the refraction of the light passing through it.

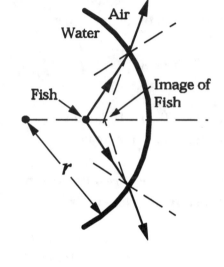

(*a*) Use the equation for refraction at a single surface to relate the image and object distances:

$$\frac{n_1}{s} + \frac{n_2}{s'} = \frac{n_2 - n_1}{r}$$

Solving for s' yields:

$$s' = \frac{n_2}{\dfrac{n_2 - n_1}{r} - \dfrac{n_1}{s}}$$

Substitute numerical values
($s = -10$ cm, $n_1 = 1.33$, $n_2 = 1.00$ and $r = 20$ cm) and evaluate s':

$$s' = \frac{1.00}{\dfrac{1.00 - 1.33}{20\,\text{cm}} - \dfrac{1.33}{-10\,\text{cm}}} = \boxed{8.6\,\text{cm}}$$

and the image is 8.6 cm from the front surface of the bowl.

(b) For s = 30 cm:

$$s' = \cfrac{1.00}{\cfrac{1.00 - 1.33}{20\text{ cm}} - \cfrac{1.33}{-30\text{ cm}}} = 35.9\text{ cm}$$

The change in the fish's apparent location is:

$$35.9\text{ cm} - 8.6\text{ cm} \approx \boxed{27\text{ cm}}$$

37 ••   Repeat Problem 34 for when the glass rod and the object are immersed in water and (a) the object is 6.00 cm from the spherical surface, and (b) the object is 12.0 cm from the spherical surface.

Picture the Problem We can use the equation for refraction at a single surface to find the images corresponding to these three object positions. The signs of the image distances will tell us whether the images are real or virtual and the ray diagrams will confirm the correctness of our analytical solutions.

Use the equation for refraction at a single surface to relate the image and object distances:

$$\frac{n_1}{s} + \frac{n_2}{s'} = \frac{n_2 - n_1}{r} \qquad (1)$$

Solving for s' yields:

$$s' = \cfrac{n_2}{\cfrac{n_2 - n_1}{r} - \cfrac{n_1}{s}}$$

(a) Substitute numerical values (s = 6.00 cm, $n_1 = 1.33$, $n_2 = 1.68$, and r = 7.20 cm) and evaluate s':

$$s' = \cfrac{1.68}{\cfrac{1.68 - 1.33}{7.20\text{ cm}} - \cfrac{1.33}{6.00\text{ cm}}} = \boxed{-9.7\text{ cm}}$$

where the negative distance tells us that the image is 9.7 cm in front of the surface and is virtual.

In the following ray diagram, the object is at P and the virtual image is at P′.

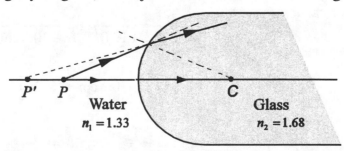

(b) Substitute numerical values (s = 12.0 cm, n_1 = 1.33, n_2 = 1.68, and r = 7.20 cm) and evaluate s':

$$s' = \frac{1.68}{\dfrac{1.68-1.33}{7.20\,\text{cm}} - \dfrac{1.33}{12.0\,\text{cm}}} = \boxed{-27\,\text{cm}}$$

where the negative distance tells us that the image is 27 cm in front of the surface and is virtual.

In the following ray diagram, the object is at P and the virtual image is at P'.

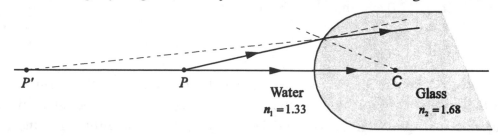

Thin Lenses

41 •   A double concave lens that has an index of refraction equal to 1.45 has radii whose magnitudes are equal to 30.0 cm and 25.0 cm. An object is located 80.0 cm to the left of the lens. Find (a) the focal length of the lens, (b) the location of the image, and (c) the magnification of the image. (d) Is the image real or virtual? Is the image upright or inverted?

Picture the Problem We can use the lens-maker's equation to find the focal length of the lens and the thin-lens equation to locate the image. We can use $m = -\dfrac{s'}{s}$ to find the lateral magnification of the image.

(a) The lens-maker's equation is:

$$\frac{1}{f} = \left(\frac{n}{n_{\text{air}}} - 1\right)\left(\frac{1}{r_1} - \frac{1}{r_2}\right)$$

where the numerals 1 and 2 denote the first and second surfaces, respectively.

Substitute numerical values to obtain:

$$\frac{1}{f} = \left(\frac{1.45}{1.00} - 1\right)\left(\frac{1}{-30.0\,\text{cm}} - \frac{1}{25.0\,\text{cm}}\right)$$

Solving for f yields:

$$f = -30.3\,\text{cm} = \boxed{-30\,\text{cm}}$$

(b) Use the thin-lens equation to relate the image and object distances:

$$\frac{1}{s} + \frac{1}{s'} = \frac{1}{f} \Rightarrow s' = \frac{fs}{s-f}$$

Substitute numerical values and evaluate s':

$$s' = \frac{(-30.3\,\text{cm})(80.0\,\text{cm})}{80.0\,\text{cm} - (-30.3\,\text{cm})} = -21.98\,\text{cm}$$
$$= -22\,\text{cm}$$

The image is 22 cm from the lens and on the same side of the lens as the object.

(c) The lateral magnification of the image is given by:

$$m = -\frac{s'}{s}$$

Substitute numerical values and evaluate m:

$$m = -\frac{-21.98\,\text{cm}}{80.0\,\text{cm}} = \boxed{0.27}$$

(d) Because $s' < 0$ and $m > 0$, the image is $\boxed{\text{virtual and upright.}}$

45 ••   (a) An object that is 3.00 cm high is placed 25.0 cm in front of a thin lens that has a power equal to 10.0 D. Draw a ray diagram to find the position and the size of the image and check your results using the thin-lens equation. (b) Repeat Part (a) if the object is placed 20.0 cm in front of the lens. (c) Repeat Part (a) for an object placed 20.0 cm in front of a thin lens that has a power equal to −10.0 D.

Picture the Problem We can find the focal length of each lens from the definition of the power P, in diopters, of a lens ($P = 1/f$). The thin-lens equation can be applied to find the image distance for each lens and the size of the image can be found from the magnification equation $m = \dfrac{y'}{y} = -\dfrac{s'}{s}$.

(a) The parallel and central rays were used to locate the image in the diagram shown below.

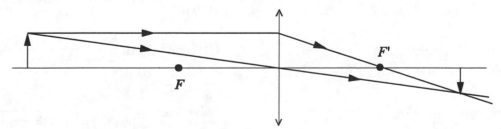

The image is real, inverted, and diminished.

Solving the thin-lens equation for s' yields:

$$\frac{1}{s} + \frac{1}{s'} = \frac{1}{f} \Rightarrow s' = \frac{fs}{s-f} \qquad (1)$$

Use the definition of the power of the lens to find its focal length:

$$f = \frac{1}{P} = \frac{1}{10.0\,\mathrm{m}^{-1}} = 0.100\,\mathrm{m} = 10.0\,\mathrm{cm}$$

Substitute numerical values and evaluate s':

$$s' = \frac{(10.0\,\mathrm{cm})(25.0\,\mathrm{cm})}{25.0\,\mathrm{cm} - 10.0\,\mathrm{cm}} = 16.7\,\mathrm{cm}$$

Use the lateral magnification equation to relate the height of the image y' to the height y of the object and the image and object distances:

$$m = \frac{y'}{y} = -\frac{s'}{s} \Rightarrow y' = -\frac{s'}{s}y \qquad (2)$$

Substitute numerical values and evaluate y':

$$y' = -\frac{16.7\,\mathrm{cm}}{25.0\,\mathrm{cm}}(3.00\,\mathrm{cm}) = -2.00\,\mathrm{cm}$$

$s' = 16.7\,\mathrm{cm}$, $y' = -2.00\,\mathrm{cm}$. Because $s' > 0$, the image is real, and because $y'/y = -0.67$ cm, the image is inverted and diminished. These results confirm those obtained graphically.

(b) The parallel and central rays were used to locate the image in the following diagram.

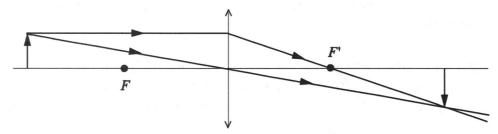

The image is real and inverted and appears to be the same size as the object.

Use the definition of the power of the lens to find its focal length:

$$f = \frac{1}{10.0\,\mathrm{m}^{-1}} = 0.100\,\mathrm{m} = 10.0\,\mathrm{cm}$$

Substitute numerical values in equation (1) and evaluate s':

$$s' = \frac{(10.0\,\mathrm{cm})(20.0\,\mathrm{cm})}{20.0\,\mathrm{cm} - 10.0\,\mathrm{cm}} = \boxed{20.0\,\mathrm{cm}}$$

Substitute numerical values in equation (2) and evaluate y':

$$y' = -\frac{20.0\,\mathrm{cm}}{20.0\,\mathrm{cm}}(3.00\,\mathrm{cm}) = \boxed{-3.00\,\mathrm{cm}}$$

Because $s' > 0$ and $y' = -3.00$ cm, the image is real, inverted, and the same size as the object. These results confirm those obtained from the ray diagram.

(c) The parallel and central rays were used to locate the image in the following diagram.

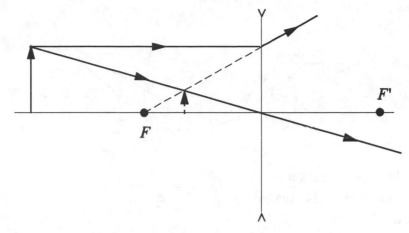

The image is virtual, upright, and diminished.

Use the definition of the power of the lens to find its focal length:

$$f = \frac{1}{-10.0\,\mathrm{m}^{-1}} = -0.100\,\mathrm{m} = -10.0\,\mathrm{cm}$$

Substitute numerical values in equation (1) and evaluate s':

$$s' = \frac{(-10.0\,\mathrm{cm})(20.0\,\mathrm{cm})}{20.0\,\mathrm{cm} - (-10.0\,\mathrm{cm})} = \boxed{-6.67\,\mathrm{cm}}$$

Substitute numerical values in equation (2) and evaluate y':

$$y' = -\frac{-6.67\,\mathrm{cm}}{20.0\,\mathrm{cm}}(3.00\,\mathrm{cm}) = \boxed{1.00\,\mathrm{cm}}$$

Because $s' < 0$ and $y' = 1.00$ cm, the image is virtual, erect, and about one-third the size of the object. These results are consistent with those obtained graphically.

49 ••   Two converging lenses, each having a focal length equal to 10 cm, are separated by 35 cm. An object is 20 cm to the left of the first lens. (a) Find the position of the final image using both a ray diagram and the thin-lens equation. (b) Is the final image real or virtual? Is the final image upright or inverted? (c) What is the overall lateral magnification of the final image?

Picture the Problem We can apply the thin-lens equation to find the image formed in the first lens and then use this image as the object for the second lens.

(*a*)   The parallel, central, and focal rays were used to locate the image formed by the first lens and the parallel and central rays to locate the image formed by the second lens.

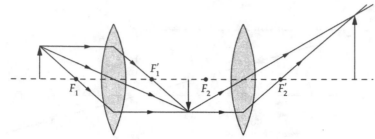

Apply the thin-lens equation to express the location of the image formed by the first lens:

$$s_1' = \frac{f_1 s_1}{s_1 - f_1} \qquad (1)$$

Substitute numerical values and evaluate s_1':

$$s_1' = \frac{(10\,\text{cm})(20\,\text{cm})}{20\,\text{cm} - 10\,\text{cm}} = 20\,\text{cm}$$

Find the lateral magnification of the first image:

$$m_1 = -\frac{s_1'}{s} = -\frac{20\,\text{cm}}{20\,\text{cm}} = -1.0$$

Because the lenses are separated by 35 cm, the object distance for the second lens is 35 cm − 20 cm = 15 cm. Equation (1) applied to the second lens is:

$$s_2' = \frac{f_2 s_2}{s_2 - f_2}$$

Substitute numerical values and evaluate s_2':

$$s_2' = \frac{(10\,\text{cm})(15\,\text{cm})}{15\,\text{cm} - 10\,\text{cm}} = 30\,\text{cm}$$

and the final image is $\boxed{85\,\text{cm}}$ to the right of the object.

Find the lateral magnification of the second image:

$$m_2 = -\frac{s_2'}{s} = -\frac{30\,\text{cm}}{15\,\text{cm}} = -2.0$$

(*b*) Because $s_2' > 0$, the image is real and because $m = m_1 m_2 = 2.0$, the image is erect and twice the size of the object.

(*c*) The overall lateral magnification of the image is the product of the magnifications of each image:

$$m = m_1 m_2 = (-1.0)(-2.0) = \boxed{2.0}$$

55 ••   An object is 15.0 cm in front of a converging lens that has a focal length equal to 15.0 cm. A diverging lens that has a focal length whose magnitude is equal to 15.0 cm is located 20.0 cm in back of the first. (*a*) Find the location of the final image and describe its properties (for example, real and inverted) and (*b*) draw a ray diagram to corroborate your answers to Part (*a*).

Picture the Problem We can apply the thin-lens equation to find the image formed in the first lens and then use this image as the object for the second lens.

Apply the thin-lens equation to express the location of the image formed by the first lens:	$$s_1' = \frac{f_1 s_1}{s_1 - f_1}$$	(1)

Substitute numerical values and evaluate s_1':

$$s_1' = \frac{(15.0\,\text{cm})(15.0\,\text{cm})}{15.0\,\text{cm} - 15.0\,\text{cm}} = \infty$$

With $s_1' = \infty$, the thin-lens equation applied to the second lens becomes:

$$\frac{1}{s_2'} = \frac{1}{f_2} \Rightarrow s_2' = f_2 = \boxed{-15.0\,\text{cm}}$$

The overall magnification of the two-lens system is the magnification of the second lens:

$$m = m_2 = -\frac{s_2'}{s_1} = -\frac{-15.0\,\text{cm}}{15.0\,\text{cm}} = \boxed{1.00}$$

The final image is 20 cm from the object, virtual, erect, and the same size as the object.

A corroborating ray diagram follows:

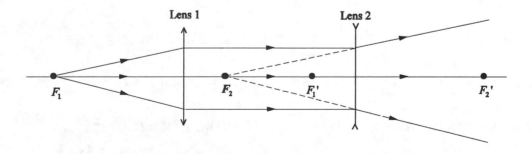

57 •••   In *Bessel's method* for finding the focal length f of a lens, an object and a screen are separated by distance L, where $L > 4f$. It is then possible to place the lens at either of two locations, both between the object and the screen, so that there is an image of the object on the screen, in one case magnified and in the other case reduced. Show that if the distance between these two lens locations is D, then the focal length is given by $f = \frac{1}{4}\left(L^2 - D^2\right)/L$. *Hint: Refer to Figure 32-60.*

Picture the Problem The ray diagram shows the two lens positions and the corresponding image and object distances (denoted by the numerals 1 and 2). We can use the thin-lens equation to relate the two sets of image and object distances to the focal length of the lens and then use the hint to express the relationships between these distances and the distances D and L to eliminate s_1, s_1', s_2, and s_2' and obtain an expression relating f, D, and L.

Relate the image and object distances for the two lens positions to the focal length of the lens:	$\dfrac{1}{s_1} + \dfrac{1}{s_1'} = \dfrac{1}{f}$ and $\dfrac{1}{s_2} + \dfrac{1}{s_2'} = \dfrac{1}{f}$
Solve for f to obtain:	$f = \dfrac{s_1 s_1'}{s_1 + s_1'} = \dfrac{s_2 s_2'}{s_2 + s_2'}$ (1)
The distances D and L can be expressed in terms of the image and object distances:	$L = s_1 + s_1' = s_2 + s_2'$ and $D = s_2 - s_1 = s_1' - s_2'$
Substitute for the sums of the image and object distances in equation (1) to obtain:	$f = \dfrac{s_1 s_1'}{L} = \dfrac{s_2 s_2'}{L}$
From the hint:	$s_1 = s_2'$ and $s_1' = s_2$
Hence $L = s_1 + s_2$ and:	$L - D = 2s_1$ and $L + D = 2s_2$
Take the product of $L - D$ and $L + D$ to obtain:	$(L - D)(L + D) = L^2 - D^2$ $= 4s_1 s_2 = 4s_1 s_1'$
From the thin-lens equation:	$4s_1 s_2 = rs_1 s_1' = 4fL$
Substitute to obtain:	$4fL = L^2 - D^2 \Rightarrow \boxed{f = \dfrac{L^2 - D^2}{4L}}$

The Eye

65 • If two point objects close together are to be seen as two distinct objects, the images must fall on the retina on two different cones that are not adjacent. That is, there must be an unactivated cone between them. The separation of the cones is about 1.00 μm. Model the eye as a uniform 2.50-cm-diameter sphere that has a refractive index of 1.34. (*a*) What is the smallest angle the two points can subtend? (See Figure 32-61.) (*b*) How close together can two points be if they are 20.0 m from the eye?

Picture the Problem We can use the relationship between a distance measured along the arc of a circle and the angle subtended at its center to approximate the smallest angle the two points can subtend and the separation of the two points 20.0 m from the eye.

(*a*) Relate θ_{min} to the diameter of the eye and the distance between the activated cones:

$$d_{eye}\theta_{min} \approx 2.00\,\mu m \Rightarrow \theta_{min} = \frac{2.00\,\mu m}{d_{eye}}$$

Substitute numerical values and evaluate θ_{min}:

$$\theta_{min} = \frac{2.00\,\mu m}{2.50\,cm} = \boxed{80.0\,\mu rad}$$

(*b*) Let D represent the separation of the points $R = 20.0$ m from the eye to obtain:

$$D = R\theta_{min} = (20.0\,m)(80.0\,\mu rad)$$
$$= \boxed{1.60\,mm}$$

67 •• The Model Eye I: A simple model for the eye is a lens that has a variable power P located a fixed distance d in front of a screen, with the space between the lens and the screen filled by air. This "eye" can focus for all values of object distance s such that $x_{np} \leq s \leq x_{fp}$ where the subscripts on the variables refer to "near point" and "far point" respectively. This "eye" is said to be normal if it can focus on very distant objects. (*a*) Show that for a normal "eye," of this type, the required minimum value of P is given by $P_{min} = 1/d$. (*b*) Show that the maximum value of P is given by $P_{max} = 1/x_{np} + 1/d$. (*c*) The difference between the maximum and minimum powers, symbolized by A, is defined as $A = P_{max} - P_{min}$ and is called the *accommodation*. Find the minimum power and accommodation for this model eye that has a screen distance of 2.50 cm, a far point distance of infinity and a near point distance of 25.0 cm.

Picture the Problem The thin-lens equation relates the image and object distances to the power of a lens.

(*a*) Use the thin-lens equation to relate the image and object distances to the power of the lens:

$$\frac{1}{s} + \frac{1}{s'} = \frac{1}{f} = P \qquad (1)$$

Because $s' = d$ and, for a distance object, $s = \infty$:

$$P_{min} = \frac{1}{s'} = \boxed{\frac{1}{d}}$$

(*b*) If x_{np} is the closest distance an object could be and still remain in clear focus on the screen, equation (1) becomes:

$$P_{max} = \boxed{\frac{1}{x_{np}} + \frac{1}{d}}$$

(*c*) Use our result in (*a*) to obtain:

$$P_{min} = \frac{1}{2.50\,cm} = \boxed{40.0\,D}$$

Use the results of (*a*) and (*b*) to express the accommodation of the model eye:

$$A = P_{max} - P_{min} = \frac{1}{x_{np}} + \frac{1}{d} - \frac{1}{d} = \frac{1}{x_{np}}$$

Substitute numerical values and evaluate A:

$$A = \frac{1}{25.0\,cm} = \boxed{4.00\,D}$$

The Simple Magnifier

75 •   What is the magnifying power of a lens that has a focal length equal to 7.0 cm when the image is viewed at infinity by a person whose near point is at 35 cm?

Picture the Problem We can use the definition of the magnifying power of a lens to find the magnifying power of this lens.

The magnifying power of the lens is given by:

$$M = \frac{x_{np}}{f}$$

Substitute numerical values and evaluate M:

$$M = \frac{35\,cm}{7.0\,cm} = \boxed{5.0}$$

77 ••   In your botany class, you examine a leaf using a convex 12-D lens as a simple magnifier. What is the angular magnification of the leaf if image formed by the lens (*a*) is at infinity and (*b*) is at 25 cm?

Picture the Problem We can use the definition of angular magnification to find the expected angular magnification if the final image is at infinity and the thin-lens equation and the expression for the magnification of a thin lens to find the angular magnification when the final image is at 25 cm.

(a) Express the angular magnification when the final image is at infinity:

$$M = \frac{x_{np}}{f} = x_{np}P$$

where P is the power of the lens.

Substitute numerical values and evaluate M:

$$M = (25\,\text{cm})(12\,\text{m}^{-1}) = \boxed{3.0}$$

(b) Express the magnification of the lens when the final image is at 25 cm:

$$m = -\frac{s'}{s}$$

Solve the thin-lens equation for s:

$$s = \frac{fs'}{s' - f}$$

Substitute for s and simplify to obtain:

$$m = -\frac{s'}{\dfrac{fs'}{s' - f}} = -\frac{s' - f}{f} = -\frac{s'}{f} + 1$$

$$= 1 - s'P$$

Substitute numerical values and evaluate m:

$$m = 1 - (-0.25\,\text{m})(12\,\text{m}^{-1}) = \boxed{4.0}$$

The Microscope

79 •• Your laboratory microscope objective has a focal length of 17.0 mm. It forms an image of a tiny specimen at 16.0 cm from its second focal point. (a) How far from the objective is the specimen located? (b) What is the magnifying power for you if your near point distance is 25.0 cm and the focal length of the eyepiece is 51.0 mm?

Picture the Problem The lateral magnification of the objective is $m_o = -L/f_o$ and the magnifying power of the microscope is $M = m_o M_e$.

(a) The lateral magnification of the objective is given by:

$$m_o = -\frac{L}{f_o}$$

| Substitute numerical values and evaluate m_o: | $m_o = -\dfrac{16\,\text{cm}}{8.5\,\text{mm}} = -18.8 = \boxed{-19}$ |

(b) The magnifying power of the microscope is given by:

$M = m_o M_e$

where M_e is the angular magnification of the lens.

| Substitute numerical values and evaluate M: | $M = (-18.8)(10) = \boxed{-1.9 \times 10^2}$ |

The Telescope

83 •   You have a simple telescope that has an objective which has a focal length of 100 cm and an eyepiece which has a focal length of 5.00 cm. You are using it to look at the moon, which subtends an angle of about 9.00 mrad. (a) What is the diameter of the image formed by the objective? (b) What angle is subtended by the image formed at infinity by the eyepiece? (c) What is the magnifying power of your telescope?

Picture the Problem Because of the great distance to the moon, its image formed by the objective lens is at the focal point of the objective lens and we can use $D = f_o \theta$ to find the diameter D of the image of the moon. Because angle subtended by the final image at infinity is given by $\theta_e = M\theta_o = M\theta$, we can solve (b) and (c) together by first using $M = -f_o/f_e$ to find the magnifying power of the telescope.

(a) Relate the diameter D of the image of the moon to the image distance and the angle subtended by the moon:

$D = s_o' \theta$

Because the image of the moon is at the focal point of the objective lens:

$s_o' = f_o$ and $D = f_o \theta$

Substitute numerical values and evaluate D:

$D = (100\,\text{cm})(9.00\,\text{mrad}) = \boxed{9.00\,\text{mm}}$

(b) and (c) Relate the angle subtended by the final image at infinity to the magnification of the telescope and the angle subtended at the objective:

$\theta_e = M\theta_o = M\theta$

Express the magnifying power of the telescope:

$$M = -\frac{f_o}{f_e}$$

Substitute numerical values and evaluate M and θ_e:

$$M = -\frac{100\,cm}{5.00\,cm} = \boxed{-20.0}$$

and

$$\theta_e = \left|-20.0\right|(9.00\,mrad) = \boxed{0.180\,rad}$$

87 •• A disadvantage of the astronomical telescope for terrestrial use (for example, at a football game) is that the image is inverted. A Galilean telescope uses a converging lens as its objective, but a diverging lens as its eyepiece. The image formed by the objective is at the second focal point of the eyepiece (the focal point on the refracted side of the eyepiece), so that the final image is virtual, upright, and at infinity. (a) Show that the magnifying power is given by $M = -f_o/f_e$, where f_o is the focal length of the objective and f_e is that of the eyepiece (which is negative). (b) Draw a ray diagram to show that the final image is indeed virtual, upright, and at infinity.

Picture the Problem The magnification of a telescope is the ratio of the angle subtended at the eyepiece lens to the angle subtended at the objective lens. We can use the geometry of the ray diagram to express both θ_e and θ_o.

(b) The ray diagram is shown below:

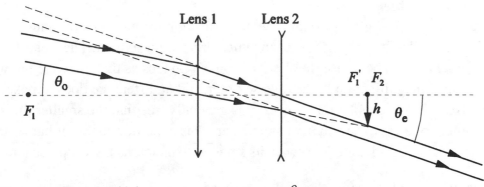

(a) Express the magnifying power M of the telescope:

$$M = \frac{\theta_e}{\theta_o}$$

Because the image formed by the objective lens is at the focal point, F'_1:

$$\theta_o = \frac{h}{f_o}$$

where we have assumed that $\theta_o \ll 1$ so that $\tan\theta_o \approx \theta_o$.

Express the angle subtended by the eyepiece:

$$\theta_e = \frac{h}{f_e} \text{ where } f_e \text{ is negative.}$$

Substitute to obtain:

$$M = \frac{\dfrac{h}{f_e}}{\dfrac{h}{f_o}} = \frac{f_o}{f_e}$$

$$\text{and } M = \boxed{-\frac{f_o}{f_e}} \text{ is positive.}$$

Remarks: Because the object for the eyepiece is at its focal point, the image is at infinity. As is also evident from the ray diagram, the image is virtual and upright.

General Problems

93 •• (*a*) Show how the same two lenses in Problem 92 should be arranged to form a compound microscope that has a tube length of 160 mm. State which lens to use as the objective, which lens to use as the eyepiece, how far apart to place the lenses, and what overall magnification you expect to get, assuming the user has a near point of 25 cm. (*b*) Draw a ray diagram to show how rays from a close object are refracted by the lenses.

Picture the Problem
Because the focal lengths appear in the magnification formula as a product, it would appear that it does not matter in which order we use them. The usual arrangement would be to use the shorter focal length lens as the objective but we get the same magnification in the reverse order. What difference does it make then? None in this problem. However, it is generally true that the smaller the focal length of a lens, the smaller its diameter. This condition makes it harder to use the shorter focal length lens, with its smaller diameter, as the eyepiece lens.

In a compound microscope, the lenses are separated by:

$$\delta = L + f_e + f_0$$

Substitute numerical values and evaluate δ:

$$\delta = 16\,\text{cm} + 7.5\,\text{cm} + 2.5\,\text{cm} = 26\,\text{cm}$$

The overall magnification of a compound microscope is given by:

$$M = m_0 M_e = -\frac{L}{f_0}\frac{x_{np}}{f_e}$$

Substitute numerical values and evaluate M:

$$M = -\left(\frac{16\,\text{cm}}{7.5\,\text{cm}}\right)\left(\frac{25\,\text{cm}}{2.5\,\text{cm}}\right) = -21$$

The lens with a focal length of 25 mm should be the objective. The two lenses should be separated by 210 mm. The angular magnification is –21.

(b) A ray diagram showing how rays from a near-by object are refracted by the lenses follows. A real and inverted image of the near-by object is formed by the objective lens at the first focal point of the eyepiece lens. The eyepiece lens forms an inverted and virtual image of this image at infinity.

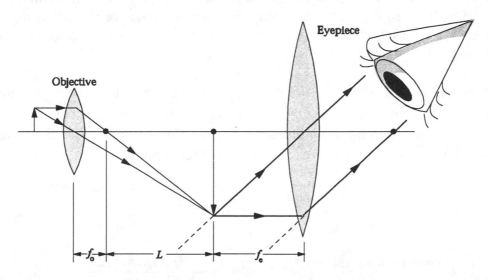

95 ••   A 35-mm digital camera has a rectangular array of CCDs (light sensors) that is 24 mm by 36 mm. It is used to take a picture of a person 175-cm tall so that the image just fills the height (24 mm) of the CCD array. How far should the person stand from the camera if the focal length of the lens is 50 mm?

Picture the Problem We can use the thin-lens equation and the definition of the magnification of an image to determine where the person should stand.

Use the thin-lens equation to relate s and s':

$$\frac{1}{s} + \frac{1}{s'} = \frac{1}{f}$$

The magnification of the image is given by:

$$m = -\frac{s'}{s} = -\frac{2.4\,\text{cm}}{175\,\text{cm}} = -1.37 \times 10^{-2}$$

and

$$s' = -ms$$

Substitute for s' to obtain:

$$\frac{1}{s} - \frac{1}{ms} = \frac{1}{f} \Rightarrow s = \left(1 - \frac{1}{m}\right)f$$

Substitute numerical values and evaluate s:

$$s = \left(1 - \frac{1}{-1.37 \times 10^{-2}}\right)(50\,\text{mm}) = \boxed{3.7\,\text{m}}$$

101 •••   An object is 15.0 cm in front of a thin converging lens that has a focal length equal to 10.0 cm. A concave mirror that has a radius equal to 10.0 cm is 25.0 cm in back of the lens. (*a*) Find the position of the final image formed by the mirror-lens combination. (*b*) Is the image real or virtual? Is the image upright or inverted? (*c*) On a diagram, show where your eye must be to see this image.

Picture the Problem We can use the thin-lens equation to locate the first image formed by the converging lens, the mirror equation to locate the image formed in the mirror, and the thin-lens equation a second time to locate the final image formed by the converging lens as the rays pass back through it.

(*a*) Solve the thin-lens equation for s_1':

$$s_1' = \frac{fs_1}{s_1 - f}$$

Substitute numerical values and evaluate s_1':

$$s_1' = \frac{(10.0\,\text{cm})(15.0\,\text{cm})}{15.0\,\text{cm} - 10.0\,\text{cm}} = 30\,\text{cm}$$

Because the image formed by the converging lens is behind the mirror:

$$s_2 = 25\,\text{cm} - 30\,\text{cm} = -5\,\text{cm}$$

Solve the mirror equation for s_2':

$$s_2' = \frac{fs_2}{s_2 - f}$$

Substitute numerical values and evaluate s_2':

$$s_2' = \frac{(5\,\text{cm})(-5\,\text{cm})}{-5\,\text{cm} - 5\,\text{cm}} = 2.50\,\text{cm} \text{ and the}$$

image is 22.5 cm from the lens; i.e., $s_3 = 22.5$ cm.

Solve the thin-lens equation for s_3':

$$s_3' = \frac{fs_3}{s_3 - f}$$

Substitute numerical values and evaluate s_3':

$$s_3' = \frac{(10.0\,\text{cm})(22.5\,\text{cm})}{22.5\,\text{cm} - 10.0\,\text{cm}} = 18\,\text{cm}$$

The final position of the image is 18 cm from the lens, on the same side as the original object.

(b) The ray diagram is shown below. The numeral 1 represents the object. The parallel and central rays from 1 are shown; one passes through the center of the converging lens, the other is paraxial and then passes through the focal point F'. The two rays intersect behind the mirror, and the image formed there, identified by the numeral 2, serves as a virtual object for the mirror. Two rays are shown emanating from this virtual image, one through the center of the mirror, the other passing through its focal point (halfway between C and the mirror surface) and then continuing as a paraxial ray. These two rays intersect in front of the mirror, forming a real image, identified by the numeral 3. Finally, the image 3 serves as a real object for the lens; again we show two rays, a paraxial ray that then passes through the focal point F and a ray through the center of the lens. These two rays intersect to form the final $\boxed{\text{real, upright,}}$ and diminished image, identified as 4.

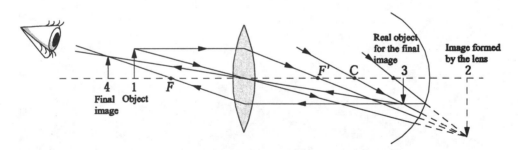

(c) To see this image the eye must be to the left of the final image.

107 •••   The lateral magnification of a spherical mirror or a thin lens is given by $m = -s'/s$. Show that for objects of small horizontal extent, the longitudinal magnification is approximately $-m^2$. *Hint:* Show that $ds'/ds = -s'^2/s^2$.

Picture the Problem Examine the amount by which the image distance s' changes due to a change in s.

Solve the thin-lens equation for s':

$$s' = \left(\frac{1}{f} - \frac{1}{s}\right)^{-1}$$

Differentiate s' with respect to s:

$$\frac{ds'}{ds} = \frac{d}{ds}\left[\left(\frac{1}{f} - \frac{1}{s}\right)^{-1}\right] = -\frac{1}{\left(\frac{1}{f} - \frac{1}{s}\right)^2}\frac{1}{s^2} = -\frac{s'^2}{s^2} = -m^2$$

The image of an object of length Δs will have a length $-m^2\Delta s$.

Chapter 33
Interference and Diffraction

Conceptual Problems

3 • The spacing between Newton's rings decreases rapidly as the diameter of the rings increases. Explain qualitatively why this occurs.

Determine the Concept The thickness of the air space between the flat glass and the lens is approximately proportional to the square of d, the diameter of the ring. Consequently, the separation between adjacent rings is proportional to $1/d$.

7 • A two-slit interference pattern is formed using monochromatic laser light with a wavelength of 640 nm. At the second maximum from the central maximum, what is the path-length difference between the light coming from each of the slits? (*a*) 640 nm (*b*) 320 nm (*c*) 960 nm (*d*) 1280 nm.

Determine the Concept For constructive interference, the path difference is an integer multiple of λ; that is, $\Delta r = m\lambda$. For $m = 2$, $\Delta r = 2(640 \text{ nm})$. $\boxed{(d)}$ is correct.

15 • True or false:

(*a*) When waves interfere destructively, the energy is converted into heat energy.
(*b*) Interference patterns are observed only if the relative phases of the waves that superimpose remain constant.
(*c*) In the Fraunhofer diffraction pattern for a single slit, the narrower the slit, the wider the central maximum of the diffraction pattern.
(*d*) A circular aperture can produce both a Fraunhofer diffraction pattern and a Fresnel diffraction pattern.
(*e*) The ability to resolve two point sources depends on the wavelength of the light.

(*a*) False. When destructive interference of light waves occurs, the energy is no longer distributed evenly. For example, light from a two-slit device forms a pattern with very bright and very dark parts. There is practically no energy at the dark fringes and a great deal of energy at the bright fringe. The total energy over the entire pattern equals the energy from one slit plus the energy from the second slit. Interference re-distributes the energy.

(*b*) True.

(c) True. The width of the central maximum in the diffraction pattern is given by $\theta_m = \sin^{-1}\dfrac{m\lambda}{a}$ where a is the width of the slit. Hence, the narrower the slit, the wider the central maximum of the diffraction pattern.

(d) True.

(e) True. The critical angle for the resolution of two sources is directly proportional to the wavelength of the light emitted by the sources ($\alpha_c = 1.22\dfrac{\lambda}{D}$).

Estimation and Approximation

19 ••   (a) Estimate how close an approaching car at night on a flat, straight stretch of highway must be before its headlights can be distinguished from the single headlight of a motorcycle. (b) Estimate how far ahead of you a car is if its two red taillights merge to look as if they were one.

Picture the Problem Assume a separation of 1.5 m between typical automobile headlights and tail lights, a nighttime pupil diameter of 5.0 mm, 550 nm for the wavelength of the light (as an average) emitted by the headlights, 640 nm for red taillights, and apply the Rayleigh criterion.

(a) The Rayleigh criterion is given by Equation 33-25:

$$\alpha_c = 1.22\frac{\lambda}{D}$$

where D is the separation of the headlights (or tail lights).

The critical angular separation is also given by:

$$\alpha = \frac{d}{L}$$

where d is the separation of head lights (or tail lights) and L is the distance to approaching or receding automobile.

Equate these expressions for α_c to obtain:

$$\frac{d}{L} = 1.22\frac{\lambda}{D} \Rightarrow L = \frac{Dd}{1.22\lambda}$$

Substitute numerical values and evaluate L:

$$L = \frac{(5.0\text{ mm})(1.5\text{ m})}{1.22(550\text{ nm})} \approx \boxed{11\text{ km}}$$

(b) For red light:

$$L = \frac{(5.0\text{ mm})(1.5\text{ m})}{1.22(640\text{ nm})} \approx \boxed{9.6\text{ km}}$$

21 •• Estimate the maximum distance a binary star system could be resolvable by the human eye. Assume the two stars are about fifty times further apart than the Earth and Sun are. Neglect atmospheric effects. (A test similar to this "eye test" was used in ancient Rome to test for eyesight acuity before entering the army. A normal eye could just barely resolve two well-known close-together stars in the sky. Anyone who could not tell there were two stars was rejected. In this case, the stars were not a binary system, but the principle is the same.)

Picture the Problem Assume that the diameter of a pupil at night is 5.0 mm and that the wavelength of light is in the middle of the visible spectrum at about 550 nm. We can use the Rayleigh criterion for the separation of two sources and the geometry of the Earth-to-binary star system to derive an expression for the distance to the binary stars.

If the distance between the binary stars is represented by d and the Earth-star distance by L, then their angular separation is given by:

$$\alpha = \frac{d}{L}$$

The critical angular separation of the two sources is given by the Rayleigh criterion:

$$\alpha_c = 1.22\frac{\lambda}{D}$$

For $\alpha = \alpha_c$:

$$\frac{d}{L} = 1.22\frac{\lambda}{D} \Rightarrow L = \frac{Dd}{1.22\lambda}$$

Substitute numerical values and evaluate L:

$$L = \frac{(5.0\text{ mm})(50)(1.5\times10^{11}\text{ m})}{1.22(550\text{ nm})}$$

$$\approx 5.59\times10^{13}\text{ km}\times\frac{1\,c\cdot\text{y}}{9.461\times10^{15}\text{ m}}$$

$$\boxed{5.9\,c\cdot\text{y}}$$

Phase Difference and Coherence

23 •• Two coherent microwave sources both produce waves of wavelength 1.50 cm. The sources are located in the $z = 0$ plane, one at $x = 0$, $y = 15.0$ cm and the other at $x = 3.00$ cm, $y = 14.0$ cm. If the sources are in phase, find the difference in phase between these two waves for a receiver located at the origin.

Picture the Problem The difference in phase depends on the path difference according to $\delta = \dfrac{\Delta r}{\lambda}2\pi$. The path difference is the difference in the distances of (0, 15.0 cm) and (3.00 cm, 14.0 cm) from the origin.

Relate a path difference Δr to a phase shift δ:

$$\delta = \frac{\Delta r}{\lambda} 2\pi$$

The path difference Δr is:

$$\Delta r = 15.0\,\text{cm} - \sqrt{(3.00\,\text{cm})^2 + (14.0\,\text{cm})^2}$$
$$= 0.682\,\text{cm}$$

Substitute numerical values and evaluate δ:

$$\delta = \left(\frac{0.682\,\text{cm}}{1.50\,\text{cm}}\right) 2\pi \approx \boxed{2.9\,\text{rad}}$$

Interference in Thin Films

25 •• The diameters of fine fibers can be accurately measured using interference patterns. Two optically flat pieces of glass of length L are arranged with the wire between them, as shown in Figure 33-40. The setup is illuminated by monochromatic light, and the resulting interference fringes are observed. Suppose that L is 20.0 cm and that yellow sodium light (wavelength of 590 nm) is used for illumination. If 19 bright fringes are seen along this 20.0-cm distance, what are the limits on the diameter of the wire? *Hint: The nineteenth fringe might not be right at the end, but you do not see a twentieth fringe at all.*

Picture the Problem The condition that one sees m fringes requires that the path difference between light reflected from the bottom surface of the top slide and the top surface of the bottom slide is an integer multiple of a wavelength of the light.

The mth fringe occurs when the path difference $2d$ equals m wavelengths:

$$2d = m\lambda \;\Rightarrow\; d = \frac{m\lambda}{2}$$

Because the nineteenth (but not the twentieth) bright fringe can be seen, the limits on d must be:

$$\left(m - \tfrac{1}{2}\right)\frac{\lambda}{2} < d < \left(m + \tfrac{1}{2}\right)\frac{\lambda}{2}$$

where $m = 19$

Substitute numerical values to obtain:

$$\left(19 - \tfrac{1}{2}\right)\frac{590\,\text{nm}}{2} < d < \left(19 + \tfrac{1}{2}\right)\frac{590\,\text{nm}}{2}$$

or

$$\boxed{5.5\,\mu\text{m} < d < 5.8\,\mu\text{m}}$$

29 •• A film of oil that has an index of refraction of 1.45 rests on an optically flat piece of glass with an index of refraction of 1.60. When illuminated by white light at normal incidence, light of wavelengths 690 nm and 460 nm is predominant in the reflected light. Determine the thickness of the oil film.

Picture the Problem Because there is a $\frac{1}{2}\lambda$ phase change due to reflection at both the air-oil and oil-glass interfaces, the condition for constructive interference is that twice the thickness of the oil film equal an integer multiple of the wavelength of light in the film.

Express the condition for constructive interference:	$2t = \lambda', 2\lambda', 3\lambda', \ldots = m\lambda' \qquad (1)$ where λ' is the wavelength of light in the oil and $m = 0, 1, 2, \ldots$
Substitute for λ' to obtain:	$2t = m\dfrac{\lambda}{n} \Rightarrow \lambda = \dfrac{2nt}{m}$ where n is the index of refraction of the oil.
Substitute for the predominant wavelengths to obtain:	$690\,\text{nm} = \dfrac{2nt}{m}$ and $460\,\text{nm} = \dfrac{2nt}{m+1}$
Divide the first of these equations by the second and simplify to obtain:	$\dfrac{690\,\text{nm}}{460\,\text{nm}} = \dfrac{\dfrac{2nt}{m}}{\dfrac{2nt}{m+1}} = \dfrac{m+1}{m} \Rightarrow m = 2$
Solve equation (1) for t:	$t = \dfrac{m\lambda}{2n}$
Substitute numerical values and evaluate t:	$t = \dfrac{(2)(690\,\text{nm})}{2(1.45)} = \boxed{476\,\text{nm}}$

Newton's Rings

31 ••   A Newton's ring apparatus consists of a plano-convex glass lens with radius of curvature R that rests on a flat glass plate, as shown in Figure 33-42. The thin film is air of variable thickness. The apparatus is illuminated from above by light from a sodium lamp that has a wavelength of 590 nm. The pattern is viewed by reflected light. (*a*) Show that for a thickness t the condition for a bright (constructive) interference ring is $2t = \left(m + \frac{1}{2}\right)\lambda$ where $m = 0, 1, 2, \ldots$
(*b*) Show that for $t \ll R$, the radius r of a fringe is related to t by $r = \sqrt{2tR}$.
(*c*) For a radius of curvature of 10.0 m and a lens diameter of 4.00 cm. How many bright fringes would you see in the reflected light? (*d*) What would be the diameter of the sixth bright fringe? (*e*) If the glass used in the apparatus has an index of refraction $n = 1.50$ and water replaces the air between the two pieces of glass, explain qualitatively the changes that will take place in the bright-fringe pattern.

Picture the Problem This arrangement is essentially identical to a "thin film" configuration, except that the "film" is air. A phase change of 180° ($\frac{1}{2}\lambda$) occurs at the top of the flat glass plate. We can use the condition for constructive interference to derive the result given in (*a*) and use the geometry of the lens on the plate to obtain the result given in (*b*). We can then use these results in the remaining parts of the problem.

(*a*) The condition for constructive interference is:

$$2t + \tfrac{1}{2}\lambda = \lambda, 2\lambda, 3\lambda, \ldots$$

or

$$2t = \tfrac{1}{2}\lambda, \tfrac{3}{2}\lambda, \tfrac{5}{2}\lambda, \ldots = \left(m + \tfrac{1}{2}\right)\lambda$$

where λ is the wavelength of light in air and $m = 0, 1, 2, \ldots$

Solving for t yields:

$$t = \boxed{\left(m + \tfrac{1}{2}\right)\frac{\lambda}{2}, m = 0, 1, 2, \ldots} \qquad (1)$$

(*b*) From Figure 33-42 we have:

$$r^2 + (R - t)^2 = R^2$$

or

$$R^2 = r^2 + R^2 - 2Rt + t^2$$

For $t \ll R$ we can neglect the last term to obtain:

$$R^2 \approx r^2 + R^2 - 2Rt \Rightarrow r = \boxed{\sqrt{2Rt}} \quad (2)$$

(*c*) Square equation (2) and substitute for t from equation (1) to obtain:

$$r^2 = \left(m + \tfrac{1}{2}\right)R\lambda \Rightarrow m = \frac{r^2}{R\lambda} - \frac{1}{2}$$

Substitute numerical values and evaluate m:

$$m = \frac{(2.00\,\text{cm})^2}{(10.0\,\text{m})(590\,\text{nm})} - \frac{1}{2} = 67$$

and so there will be $\boxed{68}$ bright fringes.

(*d*) The diameter of the mth fringe is:

$$D = 2r = 2\sqrt{\left(m + \tfrac{1}{2}\right)R\lambda}$$

Noting that $m = 5$ for the sixth fringe, substitute numerical values and evaluate D:

$$D = 2\sqrt{\left(5 + \tfrac{1}{2}\right)(10.0\,\text{m})(590\,\text{nm})}$$
$$= \boxed{1.14\,\text{cm}}$$

(*e*) The wavelength of the light in the film becomes λ_{air}/n = 444 nm. The separation between fringes is reduced (the fringes would become more closely spaced.) and the number of fringes that will be seen is increased by a factor of 1.33.

Two-Slit Interference Patterns

35 • Using a conventional two-slit apparatus with light of wavelength 589 nm, 28 bright fringes per centimeter are observed near the center of a screen 3.00 m away. What is the slit separation?

Picture the Problem We can use the expression for the distance on the screen to the mth and $(m + 1)$st bright fringes to obtain an expression for the separation Δy of the fringes as a function of the separation of the slits d. Because the number of bright fringes per unit length N is the reciprocal of Δy, we can find d from N, λ, and L.

Express the distance on the screen to the mth and $(m + 1)$st bright fringe:

$$y_m = m\frac{\lambda L}{d} \text{ and } y_{m+1} = (m+1)\frac{\lambda L}{d}$$

Subtract the first of these equations from the second to obtain:

$$\Delta y = \frac{\lambda L}{d} \Rightarrow d = \frac{\lambda L}{\Delta y}$$

Because the number of fringes per unit length N is the reciprocal of Δy:

$$d = N\lambda L$$

Substitute numerical values and evaluate d:

$$d = (28\,\text{cm}^{-1})(589\,\text{nm})(3.00\,\text{m})$$
$$= \boxed{4.95\,\text{mm}}$$

39 •• White light falls at an angle of 30° to the normal of a plane containing a pair of slits separated by 2.50 μm. What visible wavelengths give a bright interference maximum in the transmitted light in the direction normal to the plane? (See Problem 38.)

Picture the Problem Let the separation of the slits be d. We can find the total path difference when the light is incident at an angle ϕ and set this result equal to an integer multiple of the wavelength of the light to relate the angle of incidence on the slits to the direction of the transmitted light and its wavelength.

Express the total path difference:

$$\Delta \ell = d\sin\phi + d\sin\theta$$

The condition for constructive interference is:

$$\Delta \ell = m\lambda$$
where m is an integer.

Substitute to obtain:

$$d\sin\phi + d\sin\theta = m\lambda$$

| Divide both sides of the equation by d to obtain: | $$\sin\phi + \sin\theta = \frac{m\lambda}{d}$$ |

| Set $\theta = 0$ and solve for λ: | $$\lambda = \frac{d\sin\phi}{m}$$ |

| Substitute numerical values and simplify to obtain: | $$\lambda = \frac{(2.50\,\mu m)\sin 30°}{m} = \frac{1.25\,\mu m}{m}$$ |

Evaluate λ for positive integral values of m:

m	λ (nm)
1	1250
2	625
3	417
4	313

From the table we can see that 625 nm and 417 nm are in the visible portion of the electromagnetic spectrum.

Diffraction Pattern of a Single Slit

43 •••   Measuring the distance to the moon (lunar ranging) is routinely done by firing short-pulse lasers and measuring the time it takes for the pulses to reflect back from the moon. A pulse is fired from Earth. To send the pulse out, the pulse is expanded so that it fills the aperture of a 6.00-in-diameter telescope. Assuming the only thing spreading the beam out is diffraction and that the light wavelength is 500 nm, how large will the beam be when it reaches the Moon, 3.82×10^5 km away?

Picture the Problem The diagram shows the beam expanding as it travels to the moon and that portion of it that is reflected from the mirror on the moon expanding as it returns to Earth. We can express the diameter of the beam at the moon as the product of the beam divergence angle and the distance to the moon and use the equation describing diffraction at a circular aperture to find the beam divergence angle.

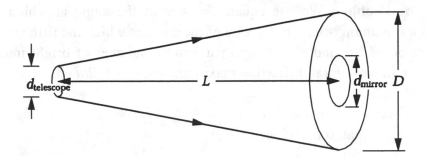

Relate the diameter D of the beam when it reaches the moon to the distance to the moon L and the beam divergence angle θ:

$$D \approx \theta L \qquad (1)$$

The angle θ subtended by the first diffraction minimum is related to the wavelength λ of the light and the diameter of the telescope opening $d_{\text{telescope}}$ by:

$$\sin \theta = 1.22 \frac{\lambda}{d_{\text{telescope}}}$$

Because $\theta \ll 1$, $\sin\theta \approx \theta$ and:

$$\theta \approx 1.22 \frac{\lambda}{d_{\text{telescope}}}$$

Substitute for θ in equation (1) to obtain:

$$D = \frac{1.22 L \lambda}{d_{\text{telescope}}}$$

Substitute numerical values and evaluate D:

$$D = \left(3.82 \times 10^8 \, \text{m}\right) \left[\frac{1.22(500 \, \text{nm})}{6.00 \, \text{in} \times \dfrac{2.54 \, \text{cm}}{\text{in}} \times \dfrac{1 \text{m}}{10^2 \, \text{cm}}} \right] = \boxed{1.53 \, \text{km}}$$

Interference-Diffraction Pattern of Two Slits

45 •• A two-slit Fraunhofer interference–diffraction pattern is observed using light that has a wavelength equal to 500 nm. The slits have a separation of 0.100 mm and an unknown width. (*a*) Find the width if the fifth interference maximum is at the same angle as the first diffraction minimum. (*b*) For that case, how many bright interference fringes will be seen in the central diffraction maximum?

Picture the Problem We can equate the sine of the angle at which the first diffraction minimum occurs to the sine of the angle at which the fifth interference maximum occurs to find a. We can then find the number of bright interference fringes seen in the central diffraction maximum using $N = 2m - 1$.

(a) Relate the angle θ_1 of the first diffraction minimum to the width a of the slits of the diffraction grating:	$\sin\theta_1 = \dfrac{\lambda}{a}$

Express the angle θ_5 corresponding to the mth fifth interference maxima maximum in terms of the separation d of the slits:	$\sin\theta_5 = \dfrac{5\lambda}{d}$

Because we require that $\theta_1 = \theta_{m5}$, we can equate these expressions to obtain:	$\dfrac{5\lambda}{d} = \dfrac{\lambda}{a} \Rightarrow a = \dfrac{d}{5}$

Substituting the numerical value of d yields:	$a = \dfrac{0.100\,\text{mm}}{5} = \boxed{20.0\,\mu\text{m}}$

(b) Because $m = 5$:	$N = 2m - 1 = 2(5) - 1 = \boxed{9}$

Using Phasors to Add Harmonic Waves

49 •   Find the resultant of the two waves whose electric fields at a given location vary with time as follows: $\vec{E}_1 = 2A_0 \sin\omega t\,\hat{i}$ and $\vec{E}_2 = 3A_0 \sin\left(\omega t + \frac{3}{2}\pi\right)\hat{i}$.

Picture the Problem Chose the coordinate system shown in the phasor diagram. We can use the standard methods of vector addition to find the resultant of the two waves.

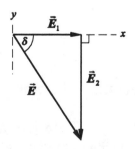

The resultant of the two waves is of the form:	$\vec{E} = \left|\vec{E}\right|\sin(\omega t + \delta)\hat{i}$	(1)

The magnitude of $\left|\vec{E}\right|$ is:	$\left|\vec{E}\right| = \sqrt{(2A_0)^2 + (3A_0)^2} = 3.6A_0$

The phase angle δ is:

$$\delta = \tan^{-1}\left(\frac{-3A_0}{2A_0}\right) = -0.98 \text{ rad}$$

Substitute for $\left|\vec{E}\right|$ and δ in equation (1) to obtain:

$$\vec{E} = \boxed{3.6A_0 \sin(\omega t - 0.98 \text{ rad})\hat{i}}$$

53 ••   Monochromatic light is incident on a sheet that has four long narrow parallel equally spaced slits a distance d apart. (*a*) Show that the positions of the interference minima on a screen a large distance L away from four equally spaced sources (spacing d, with $d \gg \lambda$) are given approximately by $y_m = m\lambda L/4d$ where $m = 1, 2, 3, 5, 6, 7, 9, 10, \ldots$ that is, m is *not* a multiple of 4. (*b*) For a screen distance of 2.00 m, light wavelength of 600 nm, and a source spacing of 0.100 mm, calculate the width of the principal interference maxima (the distance between successive minima) for four sources. Compare this width with that for two sources with the same spacing.

Picture the Problem We can use phasor concepts to find the phase angle δ in terms of the number of phasors N (four in this problem) forming a closed polygon of N sides at the minima and then use this information to express the path difference Δr for each of these locations. Applying a small angle approximation, we can obtain an expression for y that we can evaluate for enough of the path differences to establish the pattern given in the problem statement.

(*a*) Express the phase angle δ in terms of the number of phasors N forming a closed polygon of N sides:

$$\delta = m\left(\frac{2\pi}{N}\right)$$

where $m = 1, 2, 3, 4, 5, 6, ,7, \ldots$

For four equally spaced sources, the phase angle is:

$$\delta = m\left(\frac{\pi}{2}\right)$$

Express the path difference corresponding to this phase angle to obtain:

$$\Delta r = \left(\frac{\lambda}{2\pi}\right)\delta = m\frac{\lambda}{4} \qquad (1)$$

Interference maximum occur for:

$$m = 3, 6, 9, 12, \ldots$$

Interference minima occur for:

$$m = 1, 2, 4, 5, 7, 8, \ldots$$

(Note that m is not a multiple of 3.)

Express the path difference Δr in terms of $\sin\theta$ and the separation d of the slits:

$\Delta r = d\sin\theta$

or, provided the small angle approximation is valid,

$$\Delta r = \frac{yd}{L} \Rightarrow y = \frac{L}{d}\Delta r$$

Substituting for Δr from equation (1) yields:

$$\boxed{y_{min} = \frac{m\lambda L}{4d}, \quad m = 1,2,3,5,6,7,9,...}$$

(b) For $L = 2.00$ m, $\lambda = 600$ nm, $d = 0.100$ mm, and $n = 1$:

$$2y_{min} = \frac{2(600\,\text{nm})(2.00\,\text{m})}{4(0.100\,\text{mm})} = \boxed{6.00\,\text{mm}}$$

For two slits:

$$2y_{min} = \frac{2(m+\frac{1}{2})\lambda L}{d}$$

For $L = 2.00$ m, $\lambda = 600$ nm, $d = 0.100$ mm, and $m = 0$:

$$2y_{min} = \frac{(600\,\text{nm})(2.00\,\text{m})}{0.100\,\text{mm}} = 12.0\,\text{mm}$$

The width for four sources is half the width for two sources.

55 ••• Three slits, each separated from its neighbor by 60.0 μm, are illuminated at the central intensity maximum by a coherent light source of wavelength 550 nm. The slits are extremely narrow. A screen is located 2.50 m from the slits. The intensity is 50.0 mW/m². Consider a location 1.72 cm from the central maximum. (a) Draw a phasor diagram suitable for the addition of the three harmonic waves at that location. (b) From the phasor diagram, calculate the intensity of light at that location.

Picture the Problem We can find the phase constant δ from the geometry of the diagram to the right. Using the value of δ found in this fashion we can express the intensity at the point 1.72 cm from the centerline in terms of the intensity on the centerline. On the centerline, the amplitude of the resultant wave is 3 times that of each individual wave and the intensity is 9 times that of each source acting separately.

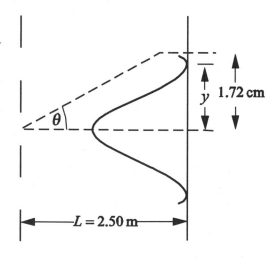

(a) Express δ for the adjacent slits:

$$\delta = \frac{2\pi}{\lambda} d \sin\theta$$

For $\theta \ll 1$, $\sin\theta \approx \tan\theta \approx \theta$:

$$\sin\theta \approx \tan\theta = \frac{y}{L}$$

Substitute to obtain:

$$\delta = \frac{2\pi d y}{\lambda L}$$

Substitute numerical values and evaluate δ:

$$\delta = \frac{2\pi(60.0\,\mu\text{m})(1.72\,\text{cm})}{(550\,\text{nm})(2.50\,\text{m})}$$

$$= \frac{3\pi}{2}\text{rad} = 270°$$

The three phasors, 270° apart, are shown in the diagram to the right. Note that they form three sides of a square. Consequently, their sum, shown as the resultant R, equals the magnitude of one of the phasors.

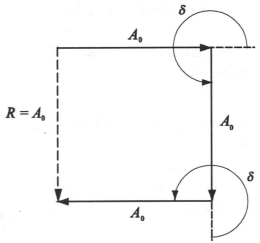

(b) Express the intensity at the point 1.72 cm from the centerline:

$$I \propto R^2$$

Because $I_0 \propto 9R^2$:

$$\frac{I}{I_0} = \frac{R^2}{9R^2} \Rightarrow I = \frac{I_0}{9}$$

Substitute for I_0 and evaluate I:

$$I = \frac{50.0\,\text{mW/m}^2}{9} = \boxed{5.56\,\text{mW/m}^2}$$

Diffraction and Resolution

57 •   Light that has a wavelength equal to 700 nm is incident on a pinhole of diameter 0.100 mm. (a) What is the angle between the central maximum and the first diffraction minimum for a Fraunhofer diffraction pattern? (b) What is the distance between the central maximum and the first diffraction minimum on a screen 8.00 m away?

Picture the Problem We can use $\theta = 1.22\dfrac{\lambda}{D}$ to find the angle between the central maximum and the first diffraction minimum for a Fraunhofer diffraction pattern and the diagram to the right to find the distance between the central maximum and the first diffraction minimum on a screen 8 m away from the pinhole.

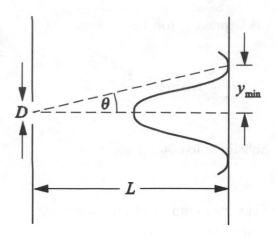

(a) The angle between the central maximum and the first diffraction minimum for a Fraunhofer diffraction pattern is given by:

$$\theta = 1.22\frac{\lambda}{D}$$

Substitute numerical values and evaluate θ:

$$\theta = 1.22\left(\frac{700\,\text{nm}}{0.100\,\text{mm}}\right) = \boxed{8.54\,\text{mrad}}$$

(b) Referring to the diagram, we see that:

$$y_{\text{min}} = L\tan\theta$$

Substitute numerical values and evaluate y_{min}:

$$y_{\text{min}} = (8.00\,\text{m})\tan(8.54\,\text{mrad})$$
$$= \boxed{6.83\,\text{cm}}$$

61 ••   The telescope on Mount Palomar has a diameter of 200 in. Suppose a double star were 4.00 light-years away. Under ideal conditions, what must be the minimum separation of the two stars for their images to be resolved using light that has a wavelength equal to 550 nm?

Picture the Problem We can use Rayleigh's criterion for circular apertures and the geometry of the diagram to obtain an expression we can solve for the minimum separation Δx of the stars.

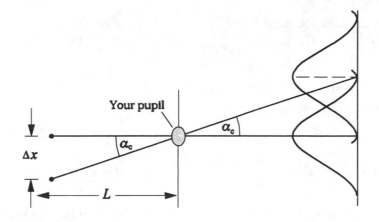

Rayleigh's criterion is satisfied provided:

$$\alpha_c = 1.22\frac{\lambda}{D}$$

Relate α_c to the separation Δx of the light sources:

$$\alpha_c \approx \frac{\Delta x}{L} \text{ because } \alpha_c << 1$$

Equate these expressions to obtain:

$$\frac{\Delta x}{L} = 1.22\frac{\lambda}{D} \Rightarrow \Delta x = 1.22\frac{\lambda L}{D}$$

Substitute numerical values and evaluate Δx:

$$\Delta x = 1.22 \left(\frac{(550\,\text{nm})\left(4\,c \cdot y \times \dfrac{9.461 \times 10^{15}\,\text{m}}{1\,c \cdot y} \right)}{200\,\text{in} \times \dfrac{2.54\,\text{cm}}{1\,\text{in}}} \right) = \boxed{5.00 \times 10^{9}\,\text{m}}$$

Diffraction Gratings

63 •   A diffraction grating that has 2000 slits per centimeter is used to measure the wavelengths emitted by hydrogen gas. (*a*) At what angles in the first-order spectrum would you expect to find the two violet lines that have wavelengths of 434 nm and 410 nm? (*b*) What are the angles if the grating has 15 000 slits per centimeter?

Picture the Problem We can solve $d \sin\theta = m\lambda$ for θ with $m = 1$ to express the location of the first-order maximum as a function of the wavelength of the light.

(*a*) The interference maxima in a diffraction pattern are at angles θ given by:

$d \sin\theta = m\lambda$
where d is the separation of the slits and $m = 0, 1, 2, \dots$

Solve for the angular location θ_m of the maxima :	$\theta_m = \sin^{-1}\left(\dfrac{m\lambda}{d}\right)$

Relate the number of slits N per centimeter to the separation d of the slits:	$N = \dfrac{1}{d}$

Substitute for d to obtain:	$\theta_m = \sin^{-1}(mN\lambda)$ $\qquad$ (1)

Evaluate θ for $\lambda = 434$ nm and $m = 1$:	$\theta_1 = \sin^{-1}\left[(2000\,\mathrm{cm}^{-1})(434\,\mathrm{nm})\right]$ $= \boxed{86.9\,\mathrm{mrad}}$

Evaluate θ for $\lambda = 410$ nm and $m = 1$:	$\theta_1 = \sin^{-1}\left[(2000\,\mathrm{cm}^{-1})(410\,\mathrm{nm})\right]$ $= \boxed{82.1\,\mathrm{mrad}}$

(b) Use equation (1) to evaluate θ for $\lambda = 434$ nm and $m = 1$:	$\theta_1 = \sin^{-1}\left[(15000\,\mathrm{cm}^{-1})(434\,\mathrm{nm})\right]$ $= \boxed{709\,\mathrm{mrad}}$

Evaluate θ for $\lambda = 410$ nm and $m = 1$:	$\theta_1 = \sin^{-1}\left[(15000\,\mathrm{cm}^{-1})(410\,\mathrm{nm})\right]$ $= \boxed{662\,\mathrm{mrad}}$

67 •• A diffraction grating that has 4800 lines per centimeter is illuminated at normal incidence with white light (wavelength range of 400 nm to 700 nm). How many orders of spectra can one observe in the transmitted light? Do any of these orders overlap? If so, describe the overlapping regions.

Picture the Problem We can use the grating equation $d \sin\theta = m\lambda, m = 1, 2, 3, \ldots$ to express the order number in terms of the slit separation d, the wavelength of the light λ, and the angle θ.

The interference maxima in the diffraction pattern are at angles θ given by:	$d \sin\theta = m\lambda \Rightarrow m = \dfrac{d \sin\theta}{\lambda}$ where $m = 1, 2, 3, \ldots$

If one is to see the complete spectrum, it must be true that:	$\sin\theta \le 1 \Rightarrow m \le \dfrac{d}{\lambda}$

Evaluate m_{max}:

$$m_{max} = \frac{\dfrac{1}{4800\,\text{cm}^{-1}}}{\lambda_{max}} = \frac{\dfrac{1}{4800\,\text{cm}^{-1}}}{700\,\text{nm}} = 2.98$$

Because $m_{max} = 2.98$, one can see the complete spectrum only for $m = 1$ and 2.

Express the condition for overlap: $m_1 \lambda_1 \geq m_2 \lambda_2$

One can see the complete spectrum for only the first and second order spectra. That is, only for $m = 1$ and 2. Because 700 nm < 2 × 400 nm, there is no overlap of the second-order spectrum into the first-order spectrum; however, there is overlap of long wavelengths in the second order with short wavelengths in the third-order spectrum.

71 •• Mercury has several stable isotopes, among them [198]Hg and [202]Hg. The strong spectral line of mercury, at about 546.07 nm, is a composite of spectral lines from the various mercury isotopes. The wavelengths of this line for [198]Hg and [202]Hg are 546.07532 nm and 546.07355 nm, respectively. What must be the resolving power of a grating capable of resolving these two isotopic lines in the third-order spectrum? If the grating is illuminated over a 2.00-cm-wide region, what must be the number of lines per centimeter of the grating?

Picture the Problem We can use the expression for the resolving power of a grating to find the resolving power of the grating capable of resolving these two isotopic lines in the third-order spectrum. Because the total number of the slits of the grating N is related to width w of the illuminated region and the number of lines per centimeter of the grating and the resolving power R of the grating, we can use this relationship to find the number of lines per centimeter of the grating.

The resolving power of a diffraction grating is given by:

$$R = \frac{\lambda}{|\Delta\lambda|} = mN \qquad (1)$$

Substitute numerical values and evaluate R:

$$R = \frac{546.07532}{|546.07532 - 546.07355|}$$

$$= 3.0852 \times 10^5 = \boxed{3.09 \times 10^5}$$

Express n, be the number of lines per centimeter of the grating, in terms of the total number of slits N of the grating and the width w of the grating:

$$n = \frac{N}{w}$$

From equation (1) we have:

$$N = \frac{R}{m}$$

Substitute for N to obtain:

$$n = \frac{R}{mw}$$

Substitute numerical values and evaluate n:

$$n = \frac{3.0852 \times 10^5}{(3)(2.00\,\text{cm})} = \boxed{5.14 \times 10^4\,\text{cm}^{-1}}$$

73 ••• For a diffraction grating in which all the surfaces are normal to the incident radiation, most of the energy goes into the zeroth order, which is useless from a spectroscopic point of view, since in zeroth order all the wavelengths are at 0°. Therefore, modern reflection gratings have shaped, or *blazed*, grooves, as shown in Figure 33-45. This shifts the specular reflection, which contains most of the energy, from the zeroth order to some higher order. (a) Calculate the *blaze angle* ϕ_m in terms of the groove separation d, the wavelength λ, and the order number m in which specular reflection is to occur for $m = 1, 2, \ldots$ (b) Calculate the proper blaze angle for the specular reflection to occur in the second order for light of wavelength 450 nm incident on a grating with 10 000 lines per centimeter.

Picture the Problem We can use the grating equation and the geometry of the grating to derive an expression for ϕ_m in terms of the order number m, the wavelength of the light λ, and the groove separation d.

(a) Because $\theta_i = \theta_r$, application of the grating equation yields:

$$d \sin(2\theta_i) = m\lambda,$$
$$\text{where } m = 0, 1, 2, \ldots \tag{1}$$

Because ϕ and θ_i have their left and right sides mutually perpendicular:

$$\theta_i = \phi_m$$

Substitute for θ_i to obtain:

$$d \sin(2\phi_m) = m\lambda$$

Solving for ϕ_m yields:

$$\phi_m = \boxed{\tfrac{1}{2}\sin^{-1}\left(m\frac{\lambda}{d}\right)}$$

(b) For $m = 2$:

$$\phi_2 = \tfrac{1}{2}\sin^{-1}\left(2\frac{450\,\text{nm}}{\dfrac{1}{10{,}000\,\text{cm}^{-1}}}\right) = \boxed{32.1°}$$

General Problems

75 •   Naturally occurring coronas (brightly colored rings) are sometimes seen around the Moon or the Sun when viewed through a thin cloud. (Warning: When viewing a sun corona, be sure that the entire sun is blocked by the edge of a building, a tree, or a traffic pole to safeguard your eyes.) These coronas are due to diffraction of light by small water droplets in the cloud. A typical angular diameter for a coronal ring is about $10°$. From this, estimate the size of the water droplets in the cloud. Assume that the water droplets can be modeled as opaque disks with the same radius as the droplet, and that the Fraunhofer diffraction pattern from an opaque disk is the same as the pattern from an aperture of the same diameter. (This last statement is known as *Babinet's principle*.)

Picture the Problem We can use $\sin\theta = 1.22\lambda/D$ to relate the diameter D of the opaque-disk water droplets to the angular diameter θ of a coronal ring and to the wavelength of light. We'll assume a wavelength of 500 nm.

The angle θ subtended by the first diffraction minimum is related to the wavelength λ of light and the diameter D of the opaque-disk water droplet:	$\sin\theta = 1.22\dfrac{\lambda}{D}$
Because of the great distance to the cloud of water droplets, $\theta \ll 1$ and:	$\theta \approx 1.22\dfrac{\lambda}{D} \Rightarrow D = \dfrac{1.22\lambda}{\theta}$
Substitute numerical values and evaluate D:	$D = \dfrac{1.22(500\,\text{nm})}{10° \times \dfrac{\pi\,\text{rad}}{180°}} = \boxed{3.5\,\mu\text{m}}$

79 •   A long, narrow horizontal slit lies 1.00 μm above a plane mirror, which is in the horizontal plane. The interference pattern produced by the slit and its image is viewed on a screen 1.00 m from the slit. The wavelength of the light is 600 nm. (*a*) Find the distance from the mirror to the first maximum. (*b*) How many dark bands per centimeter are seen on the screen?

Picture the Problem We can apply the condition for constructive interference to find the angular position of the first maximum on the screen. Note that, due to reflection, the wave from the image is $180°$ out of phase with that from the source.

(*a*) Because $y_0 \ll L$, the distance from the mirror to the first maximum is given by:	$y_0 = L\theta_0$	(1)

Express the condition for constructive interference:	$d\sin\theta = \left(m + \tfrac{1}{2}\right)\lambda$ where $m = 0, 1, 2, \dots$

Solving for θ yields:	$\theta = \sin^{-1}\left[\left(m + \tfrac{1}{2}\right)\dfrac{\lambda}{d}\right]$

For the first maximum, $m = 0$ and:	$\theta_0 = \sin^{-1}\left[\dfrac{\lambda}{2d}\right]$

Substitute in equation (1) to obtain:	$y_0 = L\sin^{-1}\left[\dfrac{\lambda}{2d}\right]$

Because the image of the slit is as far behind the mirror's surface as the slit is in front of it, $d = 2.00\ \mu\text{m}$. Substitute numerical values and evaluate y_0:	$y_0 = (1.00\,\text{m})\sin^{-1}\left[\dfrac{600\,\text{nm}}{2(2.00\,\mu\text{m})}\right]$ $= \boxed{15.1\,\text{cm}}$

(b) The separation of the fringes on the screen is given by:	$\Delta y = \dfrac{\lambda L}{d}$

The number of dark bands per unit length is the reciprocal of the fringe separation:	$n = \dfrac{1}{\Delta y} = \dfrac{d}{\lambda L}$

Substitute numerical values and evaluate n:	$n = \dfrac{2.00\,\mu\text{m}}{(600\,\text{nm})(1.00\,\text{m})} = \boxed{3.33\ \text{m}^{-1}}$

83 •• A *Fabry–Perot interferometer* (Figure 33-47) consists of two parallel, half-silvered mirrors that face each other and are separated by a small distance a. A half-silvered mirror is one that transmits 50% of the incident intensity and reflects 50% of the incident intensity. Show that when light is incident on the interferometer at an angle of incidence θ, the transmitted light will have maximum intensity when $2a = m\lambda/\cos\theta$.

Picture the Problem The *Fabry-Perot interferometer* is shown in the figure. For constructive interference in the transmitted light the path difference must be an integral multiple of the wavelength of the light. This path difference can be found using the geometry of the interferometer.

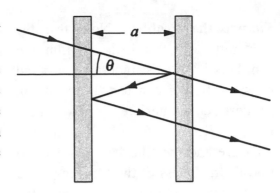

Express the path difference between the two rays that emerge from the interferometer:

$$\Delta r = \frac{2a}{\cos\theta}$$

For constructive interference we require that:

$$\Delta r = m\lambda, m = 0, 1, 2, \ldots$$

Equate these expressions to obtain:

$$m\lambda = \frac{2a}{\cos\theta}$$

Solve for $2a$ to obtain:

$$2a = \boxed{m\lambda\cos\theta}$$

85 ••   A camera lens is made of glass that has an index of refraction of 1.60. This lens is coated with a magnesium fluoride film (index of refraction 1.38) to enhance its light transmission. The purpose of this film is to produce zero reflection for light of wavelength 540 nm. Treat the lens surface as a flat plane and the film as a uniformly thick flat film. (*a*) What minimum thickness of this film will accomplish its objective? (*b*) Would there be destructive interference for any other visible wavelengths? (*c*) By what factor would the reflection for light of 400 nm wavelength be reduced by the presence of this film? Neglect the variation in the reflected light amplitudes from the two surfaces.

Picture the Problem Note that the light reflected at both the air-film and film-lens interfaces undergoes a π rad phase shift. We can use the condition for destructive interference between the light reflected from the air-film interface and the film-lens interface to find the thickness of the film. In (c) we can find the factor by which light of the given wavelengths is reduced by this film from $I \propto \cos^2 \frac{1}{2}\delta$.

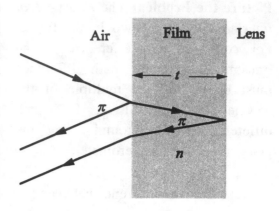

(a) Express the condition for destructive interference between the light reflected from the air-film interface and the film-lens interface:

$$2t = \left(m + \tfrac{1}{2}\right)\lambda_{film} = \left(m + \tfrac{1}{2}\right)\frac{\lambda_{air}}{n} \quad (1)$$

where $m = 0, 1, 2, \ldots$

Solving for t gives:

$$t = \left(m + \tfrac{1}{2}\right)\frac{\lambda_{air}}{2n}$$

Evaluate t for $m = 0$:

$$t = \left(\frac{1}{2}\right)\frac{540\,\text{nm}}{2(1.38)} = 97.83\,\text{nm} = \boxed{97.8\,\text{nm}}$$

(b) Solve equation (1) for λ_{air} to obtain:

$$\lambda_{air} = \frac{2tn}{m + \tfrac{1}{2}}$$

Evaluate λ_{air} for $m = 1$:

$$\lambda_{air} = \frac{2(97.8\,\text{nm})(1.38)}{1 + \tfrac{1}{2}} = 180\,\text{nm}$$

No; because 180 nm is not in the visible portion of the spectrum.

(c) Express the reduction factor f as a function of the phase difference δ between the two reflected waves:

$$f = \cos^2 \tfrac{1}{2}\delta \quad\quad (2)$$

Relate the phase difference to the path difference Δr:

$$\frac{\delta}{2\pi} = \frac{\Delta r}{\lambda_{film}} \Rightarrow \delta = 2\pi\left(\frac{\Delta r}{\lambda_{film}}\right)$$

Because $\Delta r = 2t$:

$$\delta = 2\pi\left(\frac{2t}{\lambda_{film}}\right)$$

Substitute in equation (2) to obtain:

$$f = \cos^2\left[\tfrac{1}{2}\,2\pi\left(\frac{2t}{\lambda_{\text{film}}}\right)\right] = \cos^2\left[\frac{2\pi t}{\lambda_{\text{film}}}\right]$$

$$= \cos^2\left[\frac{2\pi\,nt}{\lambda_{\text{air}}}\right]$$

Evaluate f for $\lambda = 400$ nm:

$$f_{400} = \cos^2\left[\frac{2\pi\,(1.38)(97.83\,\text{nm})}{400\,\text{nm}}\right]$$

$$= \boxed{0.273}$$

87 ••   The Impressionist painter Georges Seurat used a technique called *pointillism*, in which his paintings are composed of small, closely spaced dots of pure color, each about 2.0 mm in diameter. The illusion of the colors blending together smoothly is produced in the eye of the viewer by diffraction effects. Calculate the minimum viewing distance for this effect to work properly. Use the wavelength of visible light that requires the *greatest* distance between dots, so that you are sure the effect will work for *all* visible wavelengths. Assume the pupil of the eye has a diameter of 3.0 mm.

Picture the Problem We can use the geometry of the dots and the pupil of the eye and Rayleigh's criterion to find the greatest viewing distance that ensures that the effect will work for all visible wavelengths.

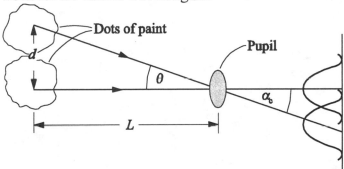

Referring to the diagram, express the angle subtended by the adjacent dots:

$$\theta \approx \frac{d}{L}$$

Letting the diameter of the pupil of the eye be D, apply Rayleigh's criterion to obtain:

$$\alpha_c = 1.22\frac{\lambda}{D}$$

Set $\theta = \alpha_c$ to obtain:

$$\frac{d}{L} = 1.22\frac{\lambda}{D} \Rightarrow L = \frac{Dd}{1.22\lambda}$$

Evaluate L for the *shortest* wavelength light in the visible portion of the spectrum:

$$L = \frac{(3.0\,\text{mm})(2.0\,\text{mm})}{(1.22)(400\,\text{nm})} = \boxed{12\,\text{m}}$$